SpringerBriefs in Applied Sciences
and Technology

Computational Intelligence

Series Editor

Janusz Kacprzyk, Systems Research Institute, Polish Academy of Sciences,
Warsaw, Poland

SpringerBriefs in Computational Intelligence are a series of slim high-quality publications encompassing the entire spectrum of Computational Intelligence. Featuring compact volumes of 50 to 125 pages (approximately 20,000-45,000 words), Briefs are shorter than a conventional book but longer than a journal article. Thus Briefs serve as timely, concise tools for students, researchers, and professionals.

More information about this subseries at http://www.springer.com/series/10618

Jagriti Saini · Maitreyee Dutta · Gonçalo Marques

Internet of Things for Indoor Air Quality Monitoring

 Springer

Jagriti Saini ⓘ
Department of Electronics
and Communication Engineering
National Institute of Technical Teachers
Training and Research
Chandigarh, India

Maitreyee Dutta ⓘ
Department of Computer Science
and Engineering
National Institute of Technical Teachers
Training and Research
Chandigarh, India

Gonçalo Marques ⓘ
ESTGOH
Polytechnic of Coimbra
Oliveira do Hospital, Portugal

ISSN 2191-530X   ISSN 2191-5318 (electronic)
SpringerBriefs in Applied Sciences and Technology
ISSN 2625-3704   ISSN 2625-3712 (electronic)
SpringerBriefs in Computational Intelligence
ISBN 978-3-030-82215-6   ISBN 978-3-030-82216-3 (eBook)
https://doi.org/10.1007/978-3-030-82216-3

This Springer imprint is published by the registered company Springer Nature Switzerland AG
The registered company address is: Gewerbestrasse 11, 6330 Cham, Switzerland

Preface

In recent years, several new application areas for indoor air quality, using emergent Internet of things technologies, have been proposed in the literature. There is currently no book that captures this growing area of research. This book aims to synthesize recent developments, present case studies, and discuss new methods in the area of knowledge. Numerous methods for air quality monitoring and analysis are available in the literature. Furthermore, the application of these methods to monitoring indoor air quality is of utmost importance for enhanced public health. In particular, the Internet of things systems can provide a continuous flow of data retrieved from cost-effective sensors that can be used in multiple applications.

This book is a merger of two relevant disciplines: computer science and air quality. The key novelty of the proposed book is to present the leading Internet of things technologies, applications, algorithms, systems, and future scope considering this multi-disciplinary domain which incorporates the data collected from smart sensors that will be processed using enhanced data analytics for enhanced indoor air quality.

The critical problem that it solves is that it presents a synthesis of the application of the Internet of things technologies in air quality data engineering. The book summarizes relevant procedures, processes, and methods on the application of the Internet of things for air quality monitoring. Moreover, this is crucial to young researchers who want to replicate these methodologies in solving their research problems on indoor air quality monitoring.

The book is structured into five chapters based on the critical areas of the application of the Internet of things for air quality monitoring. Chapters 1 and 2 will synthesize the latest knowledge in the two main sub-areas in the field. This will serve as a guide for researchers who need new topics to work on and want to identify knowledge gaps. Chapter 3 will provide a state of the art on the topic concerning the multiple components in the Internet of things for indoor air quality monitoring. Chapter 4 will introduce the integration of Internet of things with artificial intelligence for indoor air quality monitoring. Furthermore, Chap. 5 will provide future directions for work in the field of IAQ monitoring with IoT.

The proposed book is a fundamental information source for multiple groups that range from academics to industrial professionals. Furthermore, the graduate and undergraduate students from those areas will find this book a relevant source to support their cross-domain research activities that deal with air quality monitoring. Finally, this book will also provide a useful data source for professionals such as software developers and data scientists to support their industrial activities.

Chandigarh, India   Jagriti Saini
Chandigarh, India   Maitreyee Dutta
Oliveira do Hospital, Portugal   Gonçalo Marques

Contents

About the Authors

Jagriti Saini holds a Diploma in Electronics and Communication Engineering (2010) and B.Tech in Electronics and Communication Engineering (2013). She received a Master's Degree in Electronics and Communication Engineering from the National Institute of Technical Teachers Training and Research (NITTTR), Chandigarh (Panjab University), India (2017). She was awarded a Gold Medal for securing the highest percentile in the entire university during Master's Degree. She is currently pursuing her Ph.D. in Electronics and Communication Engineering from the National Institute of Technical Teachers Training and Research (NITTTR), Chandigarh (Panjab University), India. She is also receiving an INSPIRE fellowship from the Department of Science and Technology (DST), India for carrying out her research work. Her current research interests include artificial intelligence, Internet of things, environmental monitoring, indoor air quality monitoring and prediction, healthcare systems, e-health, and autonomous systems. Her Ph.D. thesis entitled "Design and Development of Intelligent Indoor Air Quality Monitoring and Prediction System—Vayuveda" is mainly focused on developing cost-effective real-time monitoring and prediction system for indoor air quality management. She published more than 15 research papers in indoor air quality monitoring and prediction systems based on the Internet of things and artificial intelligence.

Maitreyee Dutta was born in Guwahati, India. She received the B.E. degree in Electronics and Communication Engineering from Gauhati University, the M.E. degree in Electronics and Communication Engineering, and the Ph.D. degree with specialization in Image Processing from Panjab University. She is currently Professor in the Computer Science and Engineering Department, National Institute of Technical Teachers' Training and Research, Chandigarh, India. She has more than 22 years of teaching experience. Her research interests include Internet of things, security of data, IP networks, Internet, authorization, data privacy, public key encryption, pattern clustering, cloud computing, graph theory, program testing, regression analysis, sign language recognition, medical imaging, source code (software), IEC standards, ISO standards, adders, content-based retrieval, and data compression. She has more than 100 research publications in reputed journals and conferences. She completed one sponsored research project—Establishment of Cyber Security Lab—funded by the

Ministry of IT, New Delhi, Government of India, amounting Rs. 45.65 lacs. Two sponsored Project on Advanced Cyber Security Laboratory of amount 62 lacs funded by MeitY, New Delhi, and Securing Billion of Things-SEBOT funded by All India Council of Technical Education, New Delhi, of amount Rs. 14.98 lacs are in progress.

Gonçalo Marques holds a Ph.D. in Computer Science Engineering and is Member of the Portuguese Engineering Association (Ordem dos Engenheiros). He is currently working as Assistant Professor lecturing courses on programming, multimedia, and database systems. Furthermore, he worked as Software Engineer in the Innovation and Development Unit of Groupe PSA automotive industry from 2016 to 2017 and in the IBM group from 2018 to 2019. His current research interests include Internet of things, enhanced living environments, machine learning, e-health, telemedicine, medical and healthcare systems, indoor air quality monitoring and assessment, and wireless sensor networks. He has more than 80 publications in international journals and conferences, is a frequent reviewer of journals and international conferences, and is also involved in several edited book projects.

List of Figures

List of Tables

Chapter 1
Indoor Air Quality: Impact on Public Health

1.1 Introduction

Over the past few years, indoor air quality (IAQ) has received considerable attention from environmental governances, political institutions and international scientific community due to its close association with public health, comfort and wellbeing [1]. Literature reflects several studies on this topic with quantitative and qualitative analysis, evidencing the increasing mortality and morbidity rates due to Indoor Air Pollution (IAP) [2]. A large group of people in the public and private indoor environments such as gyms, schools, workplaces, and homes, spend 90% of their routine time indoors [3]. Thus, IAQ leaves a significant impact on their overall quality of life. For the most vulnerable groups including household women, children, elderly and disabled people, the risk of IAP can be greater than outdoor pollution [4].

Data provided by World Health Organization (WHO) reveal that IAP causes almost 3.8 million deaths annually [5]. There are several sources of IAP in the residential and commercial buildings such as smoking, cooking, use of consumer products, electronic machines and building materials [3]. Most harmful pollutants inside buildings include particulate matter (PM), volatile organic compounds (VOC), carbon monoxide (CO) and biological pollutants [6]. The presence of these harmful pollutants in the indoor areas is closely associated with critical health consequences, especially for a large population spending most of the time indoors. Therefore, the research on control and management of air quality has changed from outdoor environments to indoor areas. Long-term and short-term exposure to IAP are linked to a variety of diseases. Therefore, government pollution control authorities, agencies and public health experts need to take potential steps for IAQ management [7].

Although the indoor environmental conditions are highly dependent on air quality, the thermal comfort, noise and light also play an critical role in this process [8]. Moreover, three essential factors that affect IAQ by considerable level are: human activity levels inside buildings, outdoor air quality and the type of building materials used [9]. Outdoor air pollutants can easily travel inside through doors and windows while

J. Saini et al., *Internet of Things for Indoor Air Quality Monitoring*,
SpringerBriefs in Computational Intelligence,
https://doi.org/10.1007/978-3-030-82216-3_1

influencing IAQ levels. The correlation between outdoor and IAQ can be defined in terms of the ventilation rates and lifetimes of each pollutant. The IAQ levels are also affected by the design and operations of the ventilation systems [10]. Moreover, the development of IAQ monitoring systems has opened doors for measurement and control of pollutant concentrations in the building environment. In this chapter, the authors discuss the impact of IAQ on public health along with prevention and control policies for improved health and well-being of building occupants. It will also provide insights into several obsolete air quality systems while discussing the gaps and opportunities in this field.

1.2   Indoor Air Quality and Most Critical Groups

IAP is responsible for several social, environmental and health issues that adversely affect young children and women all over the world [11]. The mortality and morbidity rates due to IAP in countries with a higher level of air pollution are considerably high [12]. Nearly more than half population in the entire world live in rural areas and use solid fuels such as coal, animal dung and biomass for cooking and heating purpose [13]. The IAP related threats in such households were estimated to lead more than 2 million deaths in the year 2001. It further leads to approximately 3% of the total global burden of disease [14]. Furthermore, the emission of black carbon and carbon dioxide (CO_2) from solid fuel burning are crucial contributors to global climate change [15]. Young girls, infants and women fall in the most vulnerable groups in the rural housing areas as they spend 90% of their time indoors [16]. The inefficient cookstoves and poor combustion systems put them under the risk of serious respiratory health problems such as asthma and lung disease [17]. Moreover, persons with disability and older adults fall under the most critical groups affected by varying pollutant concentrations in the indoor environment [18].

1.2.1   IAQ in Developed and Developing Countries

The impact of IAP in developed and developing countries are equally devastating. However, the sources vary depending upon the lifestyle and routines of the people [19]. Simple biomass cookstoves are commonly used in the developing countries and they turn out to be the significant contributors of IAP [20]. Biomass combustion leads to emission of a variety of harmful pollutants such as CO, PM, chlorinated organic compounds, oxygenated organic compounds, and metal hydrocarbons depending upon the type of stoves installed in the premises [21, 22]. Rural homes that prefer conventional heating and cooking systems with the frequent use of solid fuels experience 24-h mean $PM_{2.5}$ and PM_{10} concentrations rise to 10,000 μg per m^3 [20, 23]. However, the 24-h mean concentration for the respective pollutants is recommended to be not more than 25–50 μg per m^3 by experts at WHO [23]. Commonly documented ill-effects of IAP due to solid fuel burning include chronic obstructive pulmonary disease (COPD), acute respiratory infections (ARI), pulmonary tuberculosis, infant mortality, low birth weight, cataracts, laryngeal cancer and lung cancer [3].

Stats reveal that 75% of rural and 25% of the urban population in India use solid fuels for heating and cooking needs [24]. In the year 2016, the number of deaths worldwide due to household air pollution was reported to be 4 billion [24]. The housing conditions in the urban areas and developed countries are considerably improved [25]. However, living and workspaces that stay cold, damp, and over-crowded are still a matter of concern. In addition to this, smoking is a common factor contributing to poor air quality in the indoor areas in rich communities [26]. These conditions are more problematic for children that get exposed to second-hand tobacco smoke more often. Furthermore, building materials and the frequent use of chemical-rich products in urban residential apartments or office areas is another potential cause behind poor IAQ levels [27]. The main contributors of IAP in the developed nations are chemical cleaners, deodorants, carpets, textiles, poorly maintained heating and cooling systems [28]. Moreover, the lack of natural ventilation can put a major threat to the health of building occupants in the developed nations [10]. Therefore, it is critical to find relevant solutions to deal with the decaying IAQ levels in both developed and developing countries.

1.3 Potential Pollutants and Public Health

Numerous pollutants are recognized to leave a harmful impact on human health and building comfort levels. The main focus is given to PM levels, VOCs, toxic metals, NOx, microorganisms, radon and many gaseous pollutants such as ozone (O_3), sulphur dioxide (SO_2), CO, and CO_2 [29, 30]. Figure 1.1 display potential pollutants affecting IAQ levels.

- **PM**

The close association between higher levels of PM in the indoor environment is identified since the beginning of the twentieth century. The critical incidents of smog-related hospitalization and deaths were reported from Pennsylvania (1948), Belgium (1930), London and UK (1952) [31].

PM is generally classified into three different categories: PM_{10} are the particles with less than 10 μm diameter, they are known as coarse particles. Fine particles have a diameter below 2.5 μm; therefore, they are represented as $PM_{2.5}$. Furthermore, $PM_{0.1}$ are the particles with diameters less than 0.1 μm that are widely recognized as ultrafine particles [3, 31, 32].

Several studies reveal that PM levels in the indoor environment are usually higher than the outdoor environment. The most common sources of PM in the

Fig. 1.1   Potential pollutants affecting indoor air quality

building environment include particles migrated from the outdoor areas and particles produced from several indoor activities such as residential hobbies, machine
operations, smoking, cooking and fossil fuel combustion [33].

- **CO_X**

Combustion processes such as cooking and heating are the main cause behind the
generation of CO in the indoor environment. Moreover, CO levels can also rise in the
building premises through infiltered outdoor air [34]. Common sources of CO in the
indoor areas include wood stoves, gas water heaters, furnaces, fireplaces, generators,
gas space heaters and tobacco smoke [35]. The average concentration of CO in the
indoor environment is reported to be 0.5–5 ppm. However, it can rise to 5–15 ppm
or even 30 ppm in presence of gas stoves [3].

   On the other side, CO_2 is an odorless and colorless gas. This gas plays an important
role in IAQ assessment and ventilation control [36]. As per ASHRAE standards,
indoor CO_2 concentration must be below 700 ppm to ensure better human health
[37]. If the concentration levels are increased above 3000 ppm, it can pose a serious
threat to the health of building occupants with symptoms such as concentration
difficulty, fatigue, sleepiness and headache [38].

- **SO_2**

SO_2 is a significant gas in the SO_X group and it is mainly produced due to combustion
of fossil fuels. It combines the PMs and aerosols to create complex and distinct air
in the environment [39]. Common sources of SO_2 in the indoor environment include
wood stoves, coal, kerosene heaters, tobacco smoke, oil furnaces, and vented gas

appliances [40]. The SO_2 levels in the building environment are usually small and can be easily absorbed by the surfaces. The hourly concentrations can be approximately below 20 ppb [3]. However, the repeated exposure to higher SO_2 levels can impair respiratory function [39].

- **NO_X**

Two main components belong to NO_X group and leave a considerable impact on IAQ: nitrogen dioxide (NO_2) and nitric oxide (NO) [3]. These gases can be generated due to several combustion sources but mainly heaters and cooking stoves. Few additional sources include burning of kerosene, coal, oil, gas and wood [41]. Moreover, on reaction with water, NO_2 produces nitrous acid that turns out to be a strong oxidant and potential indoor air pollutant. Buildings that are closely located to roadways can have significant influence of NO_2 levels. This air pollutant is one of the main contributors to COPD [42]. Therefore, it is imperative to minimize the exposure levels to NO_2 for the building occupants.

- **O_3**

Ozone is an oxidizing agent produced due to photochemical reaction of VOCs, NO_X and O_2 [43]. Ozone can perform a rapid reaction with numerous indoor air pollutants which further cause major damage to materials and humans. The main source of production of indoor ozone is electrical devices and outdoor atmosphere. In the office areas, indoor ozone can be originated from photocopiers, air purifying devices, disinfecting devices, and other office equipment [44]. The ozone levels in the building premises depend upon several factors including air-exchange rates, indoor emission rates, reaction between chemicals, surface removal rates and outdoor ozone levels [45]. It can generally fluctuate between 20 and 80% depending upon air exchange rate and ventilation arrangements [3, 46]. The repeated inhalation of O_3 levels can cause severe respiratory health problems and cardiovascular disease symptoms [45].

- **VOCs**

VOCs are a special type of gases containing chemicals generated from several solids and liquids [3]. Numerous studies reveal that VOC concentrations in the building environment can be almost 10 times higher as compared to outdoor air pollution [47]. Common sources of VOC include indoor chemical reactions, human activities such as smoking, cooking and building materials. The VOC concentrations in the indoor environment can be also affected by building renovations, house size, air exchange rates, window and door opening [48]. A research conducted by Huang et al. [49] shows that a 90 min cooking session can produce 50 different types of VOCs to the kitchen area. VOCs can enter the human body through dermal contact, ingestion and inhalation [3]. Low concentration exposures of VOCs are not a serious concern for building occupants; however, long term exposure can potentially cause cancer and other respiratory or skin related problems [50].

- **Biological Pollutants**

Studies reveal the presence of several types of biological pollutants in the indoor environments including house dust, cat saliva, animal dander, pollen and mites [51]. The biological allergens can be originated from house plants, animals, fungi, insects, and rodents [52]. The repeated exposure to biological pollutants and allergens can cause respiratory allergic diseases, respiratory infections, sensitization and wheezing [51]. Furthermore, exposure to viruses and bacteria in the indoor environment can also cause infectious and non-infectious adverse health problems [53].

In the modern housing arrangements, building materials, décor pieces and textiles can cause a considerable rise in the IAP levels [27]. The presence of organic and inorganic pesticides has been an important matter of concern for people that are already suffering from serious respiratory health problems. The cushioned furniture, carpets, curtains and other building materials can spoil IAQ [54]. Pesticide exposure is further linked to short-term skin irritation, eye irritation, headaches, dizziness, nausea, asthma and diabetes [55].

- **Radon and Heavy Metals**

The primary cause behind the rise of radon in the indoor premises is tap water, soil gas and building materials [56]. Furthermore, the radon concentration is also influenced by masonry materials such as brick, concrete and stone. Repeated exposure to indoor radon can increase the risk of lung cancer by 3–14% [3]. The heavy metals, on the other hand, are released from fuel consumption products, smoking, building materials and infiltered outdoor pollutants [57]. Heavy metals can enter the human body through dermal contact, inhalation and ingestion. Generally, the heavy metals such as arsenic and chromium can cause cancer; whereas, cadmium and lead are the main cause behind carcinogenic effects leading to major damage to the nervous system, slow growth development and cardiovascular effects [58].

1.4 Prevention and Control Policies

Literature shows potential evidence on significant risk to human health due to long term exposure to indoor air pollutants. Therefore, government authorities, policy-makers and environmental health experts need to find relevant ways to prevent critical consequences associated with poor air quality levels [59]. The scientific community and relevant organizations are working towards the development of IAQ standards, control policies and prevention techniques [60]. WHO and EPA play important role in providing a valuable database that can be referred to eliminate consequences of IAP while protecting building occupants [61]. The main goal is to eliminate or minimize the repeated or long-term exposure to rising pollutant concentrations in the residential and commercial premises.

The indoor environment must meet standard guidelines for IAQ as well as thermal comfort levels. The well-being of building occupants is dependent on several factors such as air temperature, air velocity, air humidity, ventilation arrangement, human clothing, particle pollutants, gaseous and particle pollutants [62]. Furthermore, the awareness of the common public regarding the use and hazardous effects of fireplaces, unvented kerosene heaters, tobacco smoke, automotive exhaust, and smoke must be increased [2]. The environmental health experts advise following regular maintenance and repair procedures for appliances and furnaces. The functional appliances and equipment can prevent the production of harmful chemical, gaseous and particle pollutants by considerable level [63]. Moreover, it is equally important to increase the ventilation arrangements on the premises.

People in rural and urban areas must be made aware of the ill-effects of IAP while guiding them about relevant preventive measures. The awareness campaigns must be conducted at all levels as per WHO guidelines to maintain IAQ [64]. The authorities need to do time to time inspection at commercial and residential spaces including hospitals, schools, offices and residential societies. The government also needs to set regulations for the building materials so that the hazardous ones can be avoided. In the rural areas of developing countries such as Peru, Bangladesh, Kenya, India, Nepal, most of the people use solid fuels such as coal, cow dung, kerosene and firewood for cooking and heating purpose [20, 65, 66]. The repeated exposure to such harmful elements is the main cause behind increased mortality and morbidity rates due to IAP in these areas [67]. The researchers and environmental health experts need to find some reliable solutions to meet cooking and heating needs of this 60 to 70% of the rural population in the developing countries [65]. The main goal must be to identify budget-friendly alternatives to meet the lifestyle requirements of the target population.

The experts are also recommending the development of smart buildings where automated or standalone air quality monitoring systems can be installed to maintain regular track of changing pollutant concentrations [68, 69]. This real-time information may help building occupants to stay aware of critical consequences. Furthermore, they can take relevant steps to improve ventilation before the pollutant concentrations cross hazardous ratings. Although it is possible to find a variety of air filters such as electronic air purifiers, heat recovery ventilators and minimum efficiency reposting value filters in the market; they cannot promise complete relief from IAP. The natural ventilation arrangements can be more useful, but this idea is again challenging in the residential societies that are located close to the roadways and industrial units. Regular monitoring and prediction of IAQ levels can be a great effort towards prevention and control of pollutant concentrations in the urban as well as rural areas [70]. These systems can provide prior estimates about rising IAP concentrations so that building occupants can take preventive actions ahead of time.

1.5   Obsolete Air Quality Systems

Measuring and maintaining air quality in the building environment has been a matter of concern since the past several decades [71]. The environmental health experts and researchers over the years have developed a variety of tools and techniques to measure IAP concentrations in the living spaces [72]. However, many of those traditional methods and tools have become obsolete these days. Generally, air pollution monitoring in the early days was performed using conventional systems that were not able to provide real-time stats [72]. Moreover, the monitored data could not be accessed online. One of the oldest techniques for monitoring air quality was Manual Air Quality Monitoring Systems or AAQMS [73]. Those systems were designed using High Volume Sampler System that could sample larger air volumes in the target areas. In these systems, the devices used to sample ambient air for a few days and then the samples were transferred to the data analysis centre manually. The reports were generated after manual computation and then relevant actions were taken by environmental management authorities. The whole process used to take 2–7 days in sample collection, manual transfer to labs, analysis and control measures [74]. Moreover, these systems were prone to manual errors.

Another invention in this field as open-path monitoring systems in which instead of following measurements at specific local points, the average concentration of the pollutants over extended measurement path was considered. These systems were designed to provide direct measurements of surrounding atmosphere instead of spending several days in sample collection. These methods followed Fourier Transform Infrared Spectroscopy principle for operation [75, 76]. In this process, infrared radiations were transmitted with the help of an array of reflectors placed at a distance of a few hundred meters with the help of a telescope. The reflected radiation was focused on a detector and the attenuation of the measured beam strength shows details about the presence of gaseous concentration in the target air. These systems were widely used at municipal, construction and industrial sites. The expensive design and complexity involved in the data handling made these systems irrelevant to modern needs.

The continuous ambient air quality monitoring stations were invented in the year 1995 [77]. They were designed to measure environmental pollution due to industrial activities. These systems were also able to forecast smog levels by using measured data. These monitoring systems were able to measure SO_2, NO_2, CO, O_3, PM_{10} and $PM_{2.5}$ pollutants along with some odour parameters in the outdoor areas. Although these systems were suitable for regulatory needs; they could not be installed everywhere due to bulky and space-hungry designs. Moreover, they also require the higher operational expense and extensive labour work for routine maintenance and repair.

Those conventional systems were based on permanent tool sizes, fixed deployment, large designs, expensive management and were not efficient in terms of speed of monitoring as well. However, those systems are not relevant to the needs of the current generation that demands smart solutions for building management. The advancements in the field of wireless sensor networks (WSN) and Internet of Things

(IoT) have opened new doors for the development of highly efficient, automated and real-time monitoring systems [78, 79]. The new-age sensors have become smaller in size and they are affordable as well. Factory calibrated designs make them a reliable choice for rural and urban IAQ management. There is no need to work on the complicated electrochemical analysis, optical measurements, semiconductor-based developments and laser scattering principles. Instead, it is possible to install fully-automatic digital sensor units for real-time measurements of gases, particles and biological allergens. These portable and continuous sensor units promise the better implementation of preventive and regulatory guidelines by keeping people about IAQ conditions on the premises.

1.6   Discussion

This chapter presented relevant information about IAQ and its impact on public health. The stats presented in the above sections show a serious threat of IAP to the people spending 90% of their routine time indoors. In general, IAQ is greatly affected by human activities in the building premises, type of building materials and the presence of chemical compounds present in the workspace or living areas. Numerous studies provide evidence for the association of IAP with critical health problems. The most critical ones include chronic respiratory infections among children that is the single biggest cause behind death rates in rural areas of developing countries. The repeated exposure to biomass combustion and tobacco smoke is further linked to respiratory disease, chronic obstructive respiratory disease and lung cancer. Therefore, it is imperative to find relevant solutions for maintaining IAQ levels depending upon the socio-economic conditions of the target groups.

To eliminate troubles associated with IAP, it is not enough to focus on the contributing factors. Instead, it is equally crucial to find potential techniques to eliminate risks. Development of IAQ monitoring and prediction systems can be a great step in this direction. The real-time data collection and analysis could help building occupants to take relevant actions for IAQ management. Furthermore, it can open new doors for the environmental management authorities, policymakers and government agencies to take relevant steps towards building management and reducing IAP levels. The main focus must be given to:

- Development of reliable ventilation arrangements for rural and urban buildings.
- Improving heating and cooking practices, especially in developing countries.
- Use of organic products and appropriate building materials.

It is equally essential to focus on the diversity of the problem so that standard solution can be developed to maintain IAQ levels in residential and commercial buildings including schools, gyms, cafeteria and hospitals. The development of cost-effective monitoring systems by utilizing advanced technologies such as WSN and IoT can open doors to better IAQ management in both developed and developing countries.

Government health professionals, policymakers and researchers need to work collectively to manage IAQ in the building premises. The efforts should not be limited to the development of monitoring systems; instead, authorities need to focus on the preventive measures against IAP. The awareness programs and frequent examination of IAQ in buildings are must to maintain air quality standards. Furthermore, the development of a monitoring system must be focused on the need for economically challenging sectors of society. The correct implementation of the control policies, preventive measures and monitoring systems could improve the IAQ conditions while preventing public health hazards.

1.7   Conclusions

Poor IAQ is closely associated with critical health consequences. Therefore, this field demands the immediate attention of researchers and authorities, especially in developing countries. It is necessary to develop a solid action plan at all levels to identify sustainable and affordable IAQ management solutions for households and office areas. Although it is possible to find several conventional measuring, monitoring and preventive measures for IAQ management, the trends must revolutionize with the changing lifestyles of the building occupants.

This chapter provides deep-insights regarding IAP and its impact on public health and well-being. The previous studies also indicated the need for collective efforts from several organizations and disciplines to manage the burden of IAP worldwide. They need to address the needs of IAQ management depending upon the evolving practices and technologies. The use of an interdisciplinary approach with a partnership of different organizations can support critical efforts while reducing the harmful impacts of IAP in the building premises.

References

1. A. Cincinelli, T. Martellini, Indoor air quality and health. Int. J. Environ. Res. Publ. Health **14**(11) (2017), Art. no. 11. https://doi.org/10.3390/ijerph14111286
2. K. Balakrishnan et al., The impact of air pollution on deaths, disease burden, and life expectancy across the states of India: the global burden of disease study 2017. Lancet Planet. Health **3**(1), e26–e39 (2019). https://doi.org/10.1016/S2542-5196(18)30261-4
3. V.V. Tran, D. Park, Y.-C. Lee, Indoor air pollution, related human diseases, and recent trends in the control and improvement of indoor air quality. Int. J. Environ. Res. Publ. Health **17**(8), Jan 2020, Art. no. 8. https://doi.org/10.3390/ijerph17082927
4. S. Capolongo, G. Settimo, Indoor air quality in healing environments: impacts of physical, chemical, and biological environmental factors on users, in *Indoor Air Quality in Healthcare Facilities*, ed. by S. Capolongo, G. Settimo, M. Gola (Springer International Publishing, Cham, 2017), pp. 1–11

5. World Health Organization (2020) Household air pollution and health, in *World Health Organization*, 17 May 2020. https://www.who.int/news-room/fact-sheets/detail/household-air-pollution-and-health. Accessed 17 May 2020

6. P. Kumar, B. Imam, Footprints of air pollution and changing environment on the sustainability of built infrastructure. Sci. Total Environ. **444**, 85–101 (2013). https://doi.org/10.1016/j.scitotenv.2012.11.056

7. D. Ekmekcioglu, S.S. Keskin, Characterization of indoor air particulate matter in selected elementary schools in Istanbul, Turkey. Indoor Built Environ. **16**(2), 169–176 (2007). https://doi.org/10.1177/1420326X07076777

8. Y. Al Horr, M. Arif, M. Katafygiotou, A. Mazroei, A. Kaushik, E. Elsarrag, Impact of indoor environmental quality on occupant well-being and comfort: a review of the literature. Int. J. Sustain. Built Environ. **5**(1), 1–11 (2016). https://doi.org/10.1016/j.ijsbe.2016.03.006

9. L. Fang, G. Clausen, P.O. Fanger, Impact of temperature and humidity on the perception of indoor air quality. Indoor Air **8**(2), 80–90 (1998). https://doi.org/10.1111/j.1600-0668.1998.t01-2-00003.x

10. S.J. Emmerich, K.Y. Teichman, A.K. Persily, Literature review on field study of ventilation and indoor air quality performance verification in high-performance commercial buildings in North America. Sci. Technol. Built Environ. **23**(7), 1159–1166 (2017). https://doi.org/10.1080/23744731.2016.1274627

11. G.A. Ayoko, P. Pluschke, O. Hutzinger (eds.), *Indoor Air Pollution* (Springer, Berlin, 2004)

12. G.C. Aye, P.E. Edoja, Effect of economic growth on CO_2 emission in developing countries: evidence from a dynamic panel threshold model. Cogent Econ. Financ. **5**(1) (2017). https://doi.org/10.1080/23322039.2017.1379239

13. N. Zhao et al., Natural gas and electricity: two perspective technologies of substituting coal-burning stoves for rural heating and cooking in Hebei Province of China. Energy Sci.Eng. **7**(1), 120–131 (2019). https://doi.org/10.1002/ese3.263

14. A. Kankaria, B. Nongkynrih, S.K. Gupta, Indoor air pollution in India: implications on health and its control. Indian Commun. Med. **39**(4), 203–207 (2014). https://doi.org/10.4103/0970-0218.143019

15. N. Abas, N. Khan, Carbon conundrum, climate change, CO_2 capture and consumptions, J. CO_2 Utiliz. **8**, 39–48 (2014). https://doi.org/10.1016/j.jcou.2014.06.005

16. A. Agarwal et al., Household air pollution is associated with altered cardiac function among women in Kenya. Am. J. Respir. Crit. Care Med. **197**(7), 958–961 (2018). https://doi.org/10.1164/rccm.201704-0832LE

17. K. Rumchev, Y. Zhao, J. Spickett, Health risk assessment of indoor air quality, socioeconomic and house characteristics on respiratory health among women and children of Tirupur, South India. IJERPH **14**(4), 429 (2017). https://doi.org/10.3390/ijerph14040429

18. Y. Wu, T. Liu, S. Ling, J. Szymanski, W. Zhang, S. Su, Air quality monitoring for vulnerable groups in residential environments using a multiple hazard gas detector. Sensors **19**(2), 362 (2019). https://doi.org/10.3390/s19020362

19. J.M. Seguel, R. Merrill, D. Seguel, A.C. Campagna, Indoor air quality. Am. J. Lifestyle Med. **11**(4), 284–295 (2017). https://doi.org/10.1177/1559827616653343

20. M.M. Kelp et al., Real-time indoor measurement of health and climate-relevant air pollution concentrations during a carbon-finance-approved cookstove intervention in rural India. Dev. Eng. **3**, 125–132 (2018). https://doi.org/10.1016/j.deveng.2018.05.001

21. J. Chen et al., A review of biomass burning: emissions and impacts on air quality, health and climate in China. Sci. Total Environ. **579**, 1000–1034 (2017). https://doi.org/10.1016/j.scitotenv.2016.11.025

22. F. Petracchini et al., Influence of transport from urban sources and domestic biomass combustion on the air quality of a mountain area. Environ. Sci. Pollut. Res. **24**(5), 4741–4754 (2017). https://doi.org/10.1007/s11356-016-8111-1

23. Ambient (Outdoor) Air Pollution. https://www.who.int/news-room/fact-sheets/detail/ambient-(outdoor)-air-quality-and-health. Accessed 02 Dec 2020

24. M.A. Faizan, R. Thakur, Measuring the impact of household energy consumption on respiratory diseases in India. Glob. Health Res. Policy **4**(1), 10 (2019). https://doi.org/10.1186/s41256-019-0101-7

25. M.S. Andargie, M. Touchie, W. O'Brien, A review of factors affecting occupant comfort in multi-unit residential buildings. Build. Environ. **160**, 106182 (2019). https://doi.org/10.1016/j.buildenv.2019.106182

26. K. Slezakova, D. Castro, C. Delerue-Matos, S. Morais, M. do Carmo Pereira, Levels and risks of particulate-bound PAHs in indoor air influenced by tobacco smoke: a field measurement. Environ. Sci. Pollut. Res. **21**(6), 4492–4501 (2014). https://doi.org/10.1007/s11356-013-2391-5

27. E. Uhde, T. Salthammer, Impact of reaction products from building materials and furnishings on indoor air quality—a review of recent advances in indoor chemistry. Atmos. Environ. **41**(15), 3111–3128 (2007). https://doi.org/10.1016/j.atmosenv.2006.05.082

28. S. Nehr, E. Hösen, S. Tanabe, Emerging developments in the standardized chemical characterization of indoor air quality. Environ. Int. **98**, 233–237 (2017). https://doi.org/10.1016/j.envint.2016.09.020

29. D. Kaden, C. Mandin, G. Nielsen, P. Wolkoff, *WHO Guidelines for Indoor Air Quality: Selected Pollutants* (2010). https://www.ncbi.nlm.nih.gov/books/NBK138711/. Accessed 04 July 2019

30. E.T. Gall, E.M. Carter, C. Matt Earnest, B. Stephens, Indoor air pollution in developing countries: research and implementation needs for improvements in global public health. Am. J. Publ. Health **103**(4), e67–e72 (2013). https://doi.org/10.2105/AJPH.2012.300955

31. R.B. Hamanaka, G.M. Mutlu, Particulate matter air pollution: effects on the cardiovascular system. Front. Endocrinol. **9** (2018). https://doi.org/10.3389/fendo.2018.00680

32. B.D. Horne et al., Short-term elevation of fine particulate matter air pollution and acute lower respiratory infection. Am. J. Respir. Crit. Care Med. **198**(6), 759–766 (2018). https://doi.org/10.1164/rccm.201709-1883OC

33. M.R. Miller, C.A. Shaw, J.P. Langrish, From particles to patients: oxidative stress and the cardiovascular effects of air pollution. Future Cardiol. **8**(4), 577–602 (2012). https://doi.org/10.2217/fca.12.43

34. J.A. Raub, M. Mathieu-Nolf, N.B. Hampson, S.R. Thom, Carbon monoxide poisoning—a public health perspective. Toxicology **145**(1), 1–14 (2000). https://doi.org/10.1016/S0300-483X(99)00217-6

35. B. Ritz, F. Yu, The effect of ambient carbon monoxide on low birth weight among children born in southern California between 1989 and 1993. Environ. Health Perspect. **107**(1), 17–25 (1999)

36. X. Zhang, P. Wargocki, Z. Lian, C. Thyregod, Effects of exposure to carbon dioxide and bioeffluents on perceived air quality, self-assessed acute health symptoms, and cognitive performance. Indoor Air **27**(1), 47–64 (2017). https://doi.org/10.1111/ina.12284

37. S.J. Emmerich, A.K. Persily, State-of-the-art review of CO_2 demand controlled ventilation technology and application. National Institute of Standards and Technology, Gaithersburg, MD, NIST IR 6729 (2001). https://doi.org/10.6028/NIST.IR.6729

38. A. Persily, L. de Jonge, Carbon dioxide generation rates for building occupants. Indoor Air **27**(5), 868–879 (2017). https://doi.org/10.1111/ina.12383

39. K. Katsouyanni et al., Short term effects of ambient sulphur dioxide and particulate matter on mortality in 12 European cities: results from time series data from the APHEA project. BMJ **314**(7095), 1658 (1997). https://doi.org/10.1136/bmj.314.7095.1658

40. W.J. Seow et al., Indoor concentrations of nitrogen dioxide and sulfur dioxide from burning solid fuels for cooking and heating in Yunnan Province, China. Indoor Air **26**(5), 776–783 (2016). https://doi.org/10.1111/ina.12251

41. A. Aggarwal, D. Toshniwal, Detection of anomalous nitrogen dioxide (NO_2) concentration in urban air of India using proximity and clustering methods. J. Air Waste Manag. Assoc. **69**(7), 805–822 (2019). https://doi.org/10.1080/10962247.2019.1577314

42. Y. Kodama et al., Environmental NO_2 concentration and exposure in daily life along main roads in Tokyo. Environ. Res. **89**(3), 236–244 (2002). https://doi.org/10.1006/enrs.2002.4350

43. H. Salonen, T. Salthammer, L. Morawska, Human exposure to ozone in school and office indoor environments. Environ. Int. **119**, 503–514 (2018). https://doi.org/10.1016/j.envint.2018.07.012

44. Q. Zhang, P.L. Jenkins, Evaluation of ozone emissions and exposures from consumer products and home appliances. Indoor Air **27**(2), 386–397 (2017). https://doi.org/10.1111/ina.12307

45. C. Guo, Z. Gao, J. Shen, Emission rates of indoor ozone emission devices: a literature review. Build. Environ. **158**, 302–318 (2019). https://doi.org/10.1016/j.buildenv.2019.05.024

46. M. Braik, A. Sheta, H. Al-Hiary, Hybrid neural network models for forecasting ozone and particulate matter concentrations in the Republic of China. Air Qual. Atmos Health **13**(7):839–851 (2020). https://doi.org/10.1007/s11869-020-00841-7

47. K. Lee et al., Indoor levels of volatile organic compounds and formaldehyde from emission sources at elderly care centers in Korea. PLoS ONE **13**(6), e0197495 (2018). https://doi.org/10.1371/journal.pone.0197495

48. X.M. Wu et al., Exposures to volatile organic compounds (VOCs) and associated health risks of socio-economically disadvantaged population in a 'hot spot' in Camden, New Jersey. Atmos. Environ. (1994) **57**, 72–79 (2012). https://doi.org/10.1016/j.atmosenv.2012.04.029

49. Y. Huang, S.S.H. Ho, K.F. Ho, S.C. Lee, J.Z. Yu, P.K.K. Louie, Characteristics and health impacts of VOCs and carbonyls associated with residential cooking activities in Hong Kong. J. Hazard. Mater. **186**(1), 344–351 (2011). https://doi.org/10.1016/j.jhazmat.2010.11.003

50. X. Tang, P.K. Misztal, W.W. Nazaroff, A.H. Goldstein, Siloxanes are the most abundant volatile organic compound emitted from engineering students in a classroom. Environ. Sci. Technol. Lett. **2**(11), 303–307 (2015). https://doi.org/10.1021/acs.estlett.5b00256

51. World Health Organization. Regional Office for Europe, in *Indoor Air Quality: Biological Contaminants: Report on a WHO Meeting, Rautavaara, 29 August–2 September 1988* (World Health Organization. Regional Office for Europe, 1990)

52. M. Hulin, M. Simoni, G. Viegi, I. Annesi-Maesano, Respiratory health and indoor air pollutants based on quantitative exposure assessments. Eur. Respir. J. **40**(4), 1033–1045 (2012). https://doi.org/10.1183/09031936.00159011

53. S. Baldacci et al., Allergy and asthma: effects of the exposure to particulate matter and biological allergens. Respir. Med. **109**(9), 1089–1104 (2015). https://doi.org/10.1016/j.rmed.2015.05.017

54. J.S. Colt et al., Comparison of pesticide levels in carpet dust and self-reported pest treatment practices in four US sites. J. Exposure Sci. Environ. Epidemiol. **14**(1) (2004), Art. no. 1. https://doi.org/10.1038/sj.jea.7500307

55. K.-H. Kim, E. Kabir, S.A. Jahan, Exposure to pesticides and the associated human health effects. Sci. Total Environ. **575**, 525–535 (2017). https://doi.org/10.1016/j.scitotenv.2016.09.009

56. R.C. Bruno, Sources of indoor radon in houses: a review. J. Air Pollut. Control Assoc. **33**(2), 105–109 (1983). https://doi.org/10.1080/00022470.1983.10465550

57. M.A. Al-Rajhi, M.R.D. Seaward, A.S. Al-Aamer, Metal levels in indoor and outdoor dust in Riyadh, Saudi Arabia. Environ. Int. **22**(3), 315–324 (1996). https://doi.org/10.1016/0160-4120(96)00017-7

58. Y. Faiz, M. Tufail, M.T. Javed, M.M. Chaudhry, Naila-Siddique, Road dust pollution of Cd, Cu, Ni, Pb and Zn along Islamabad expressway, Pakistan. Microchem. J. **92**(2), 186–192 (2009). https://doi.org/10.1016/j.microc.2009.03.009

59. W.-T. Tsai, Overview of green building material (GBM) policies and guidelines with relevance to indoor air quality management in Taiwan. Environments **5**(1), 4 (2017). https://doi.org/10.3390/environments5010004

60. A. Persily, Challenges in developing ventilation and indoor air quality standards: the story of ASHRAE standard 62. Build. Environ. **91**, 61–69 (2015). https://doi.org/10.1016/j.buildenv.2015.02.026

61. Z. Argunhan, A.S. Avci, Statistical evaluation of indoor air quality parameters in classrooms of a university. Adv. Meteorol. **2018**, 1–10 (2018). https://doi.org/10.1155/2018/4391579

62. L. Zhao, W. Wu, S. Li, Design and implementation of an IoT-based indoor air quality detector with multiple communication interfaces. IEEE Internet Things J. **6**(6), 9621–9632 (2019). https://doi.org/10.1109/JIOT.2019.2930191

63. M. Irfan, M.P. Cameron, G. Hassan, Interventions to mitigate indoor air pollution: a cost-benefit analysis, in *Working Papers in Economics 18/14, University of Waikato, 2018*. Accessed: 16 Apr 2019 [Online]. Available: https://ideas.repec.org/cgi-bin/refs.cgi

64. A. Dimitriou, V. Christidou, Causes and consequences of air pollution and environmental injustice as critical issues for science and environmental education. Impact Air Pollut. Health Econ. Environ. Agric. Sources (2011). https://doi.org/10.5772/17654

65. A. Agarwal et al., Household air pollution is associated with altered cardiac function among women in Kenya. Am. J. Respir. Crit. Care Med. **197**(7), 958–961 (2017). https://doi.org/10.1164/rccm.201704-0832LE

66. D. Sharma, S. Jain, Impact of intervention of biomass cookstove technologies and kitchen characteristics on indoor air quality and human exposure in rural settings of India. Environ. Int. **123**, 240–255 (2019). https://doi.org/10.1016/j.envint.2018.11.059

67. S.S. Mitter et al., Household fuel use and cardiovascular disease mortality: Golestan cohort study. Circulation **133**(24), 2360–2369 (2016). https://doi.org/10.1161/CIRCULATIONAHA.115.020288

68. K. Akkaya, I. Guvenc, R. Aygun, N. Pala, A. Kadri, IoT-based occupancy monitoring techniques for energy-efficient smart buildings, in *2015 IEEE Wireless Communications and Networking Conference Workshops* (*WCNCW*), New Orleans, LA, USA, Mar 2015, pp. 58–63. https://doi.org/10.1109/WCNCW.2015.7122529.

69. A. Schieweck et al., Smart homes and the control of indoor air quality. Renew. Sustain. Energy Rev. **94**, 705–718 (2018). https://doi.org/10.1016/j.rser.2018.05.057

70. S.S. Ahmad, R. Urooj, M. Nawaz, Air pollution monitoring and prediction, in *Current Air Quality Issues*, ed. by F. Nejadkoorki (InTech, 2015)

71. A. Norhidayah, L. Chia-Kuang, M.K. Azhar, S. Nurulwahida, Indoor air quality and sick building syndrome in three selected buildings. Proc. Eng. **53**, 93–98 (2013). https://doi.org/10.1016/j.proeng.2013.02.014

72. P. Babu, G. Suthar, Indoor air quality and thermal comfort in green building: a study for measurement, problem and solution strategies, in *Indoor Environmental Quality*, vol. 60, ed. by A. Sharma, R. Goyal, R. Mittal (Springer Singapore, Singapore, 2020), pp. 139–146

73. Y.S. Koo, D. Choi, H.Y. Kwon, J. Han, Inverse modeling to improve emission inventory for PM_{10} forecasting in East Asia region focusing on Korea. AGU Fall Meet. Abstr. **13**, A13C-3184 (2014)

74. F. Yip et al., Assessment of traditional and improved stove use on household air pollution and personal exposures in rural western Kenya. Environ. Int. **99**, 185–191 (2017). https://doi.org/10.1016/j.envint.2016.11.015

75. R. Haus et al., Mobile Fourier-transform infrared spectroscopy monitoring of air pollution. Appl. Opt., AO **33**(24), 5682–5689 (1994). https://doi.org/10.1364/AO.33.005682

76. Trace pollutant concentrations in a multiday smog episode in the California South Coast Air Basin by long path length Fourier transform infrared spectroscopy. Environ. Sci. Technol. https://pubs.acs.org/doi/pdf/10.1021/es00092a014. Accessed 04 Jan 2021

77. J.C. Chow, Measurement methods to determine compliance with ambient air quality standards for suspended particles. J. Air Waste Manag. Assoc. **45**(5), 320–382 (1995). https://doi.org/10.1080/10473289.1995.10467369

78. H. Chojer, P.T.B.S. Branco, F.G. Martins, M.C.M. Alvim-Ferraz, S.I.V. Sousa, Development of low-cost indoor air quality monitoring devices: recent advancements. Sci. Total Environ. **727**, 138385 (2020). https://doi.org/10.1016/j.scitotenv.2020.138385

79. F. Karagulian et al., Review of the performance of low-cost sensors for air quality monitoring. Atmosphere **10**(9), 506 (2019). https://doi.org/10.3390/atmos10090506

Chapter 2
Internet of Things (IoT): The Futuristic Technology

2.1 Introduction

Internet of Things (IoT) is a widely discussed topic in the field of communication technology. This technology can connect unlimited numbers of devices and sensors to influence the way we work and live. IoT finds several applications in real-world scenarios while automating things for enhanced productivity [1]. The recorded information can be further accessed by public health experts, government authorities, and policymakers. Environment monitoring is a critical application of IoT and it further contributes to enhanced public health and well-being [2]. IoT sensors can be used to collect information about air quality pollutants from the target environment and those details can be further analyzed by expert systems to understand the indoor air quality (IAQ) conditions. Literature provides evidence for the use of IoT in IAQ monitoring, management, and control [3].

IoT architectures involve the connectivity of multiple physical objects to support sensing capabilities [4]. These objects are accessed via unique addressing mechanisms following cooperation and interaction features. IoT architectures support numerous devices including sensors, wearables, microcontrollers, actuators, and smartphones [5]. An efficient IoT architecture has to be a combination of hardware devices, open-source platforms, and enhanced software solutions for data consulting, analytics, management, and storage [6]. As IoT architectures interface with the humans on the other end, they must be designed to cooperate and contribute to a comfortable lifestyle [7, 8]. Moreover, IoT structures are aware of the human context and they play an essential role in improving our routine activity levels. IoT systems can be further powered by cloud applications to store and transfer data on a real-time basis. Therefore, IoT architectures find their way to numerous applications such as smart home development, smart medical equipment, and remote monitoring [9]. Figure 2.1 demonstrates general IoT architecture.

© The Author(s), under exclusive license to Springer Nature Switzerland AG 2021
J. Saini et al., *Internet of Things for Indoor Air Quality Monitoring*,
SpringerBriefs in Computational Intelligence,
https://doi.org/10.1007/978-3-030-82216-3_2

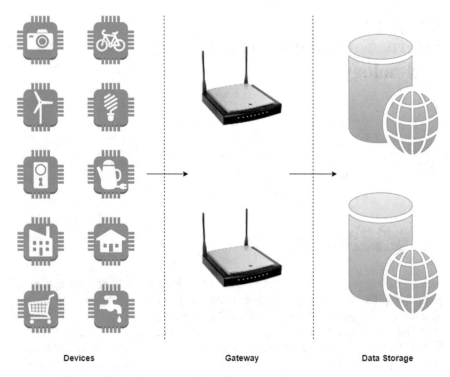

Devices   Gateway   Data Storage

Fig. 2.1 General IoT architecture

Numerous industry and academic experts have already published studies demonstrating protocols for IoT interoperability using different data structures, networks, and heterogeneous devices [3, 10]. This chapter aims to provide an introduction to IoT and its use cases in the healthcare sector. The further sections will provide details about opportunities for IoT technology in the enhanced public health, IoT open-source platforms, challenges, applications, and future directions. Finally, this chapter also synthesizes the existing literature in the field of IoT and IAQ while highlighting the potential for progress and developments in the future.

2.2 IoT for Enhanced Public Health

Currently, there are several categories of IAQ monitoring systems that make use of numerous technologies to monitor environmental parameters and human health status [11]. These systems are mainly based on advanced wireless communication technologies such as Bluetooth, ZigBee, 3G, Wi-Fi, and Ethernet [12]. The ability to monitor the physiological state of the patient provides enhanced perception about his medical state, especially for the at-risk individuals. However, monitoring of the environmental conditions plays an essential role in public health and well-being while preventing the critical consequences associated with poor air quality. The IAQ assessment is an important area of work as people spend 80–90% of their routine time indoors either at home or office [13]. Real-time 24×7 h monitoring and detection of IAQ levels can enhance the overall productivity of individuals [10, 14]. IoT is one of the most trusted technologies to design potential healthcare systems with automated features.

Despite all the advantages associated with IoT systems, there are plenty of open issues associated with the safety, confidentiality, and complexity of patient care devices. Multiple challenges in this field include network setup, normalization, data security, business models, and real-time implementation of systems [15]. Several studies support the use of wireless technologies, wearable sensors, and open-source platforms. Numerous expert systems have been already designed by existing researchers by incorporating wireless communication technologies along with mobile devices and sensors [15, 16]. However, many areas are still untouched and demand the attention of future researchers to grab new opportunities [2].

2.2.1 IoT Visions

The paradigm of IoT for IAQ monitoring applications involves three main parts: Master, Server, and Things. Master represents the build managers who are allowed to interact with the system through end-devices such as a tablet, laptop, or smartphone. Server works like a central part of the IAQ monitoring system while taking responsibility for database management, prescription generation, subsystem construction, and extensive knowledgebase management. Furthermore, Things refer to the physical objects that can be sensors for IAQ monitoring. These things are connected via multimedia technology, short message service, or wireless access network.

Based on the design and the technologies used to connect the devices in the IoT based IAQ systems, there can be three different point of views: Semantic-oriented vision, internet-oriented vision, and things-oriented vision. The semantically-oriented vision is related to the universal network of several unified objects that support searches, storage, and are capable enough to organize information. The things-oriented vision is further connected to the intelligent autonomous things that are a part of our daily lifestyle. The internet-oriented vision is mainly about the systems that are linked to the network using some unique IP addresses associated with standard protocols. When we talk about IAQ monitoring systems, the things-oriented vision is connected to the identification of high-diversity objects such as actuators and sensors. The internet-oriented vision can be related to the procedures and methods that satisfy specific data transmission requirements for IAQ monitoring systems. As these applications require a higher number of sensors to collect the massive amount of IAQ data, the architecture must be planned carefully to handle information more adequately [17]. The semantic-oriented vision connects to the methods used for processing the extensive amount of data and then extract relevant information for supporting the analysis of monitoring activities.

2.2.2   IoT Elements

There are three main elements of IoT systems: communication, sensing, and identification. The IAQ monitoring systems require proper identification so that they can meet the specific demands of the end-users. Moreover, it is not just about the physical objects; instead, all interaction entities must be acknowledged carefully to ensure adequate operation and composition of the system. Identification also plays a critical role in communication to address the items, locations, data, users, and services. Unique identifiers are applied to the systems to ensure easy identification and it ensures the unique identity of every entry in the particular application.

Literature provides evidence for multiple addressing techniques such as 6LoWPAN, IPv6, and IPv4 [18]. The sensing process includes capturing information from the target environment where the IoT system is deployed. This data can be saved at a local, remote, or cloud database depending upon the future access requirements. There is a wide range of communication protocols that can be used for executing IoT applications. The operations usually vary depending upon the data transmission range and battery dependability of the system. IoT devices make use of message queuing telemetry transport (MQTT) protocols. However, others are using a constrained application protocol (CoAP). Both of these standards support asynchronous communication mechanisms. MQTT is widely recognized as a subscribe/publish message protocol that works effectively for the lightweight machine to machine communications. Whereas, CoAP is a web transfer protocol that can be used with constrained networks and constrained nodes to handle machine to machine applications. IEEE also introduced the 802.15.4e standard in the year 2012 to complement and improve the previous 802.15.4 standard [19]. It can address the emerging needs of industrial and manufacturing applications. Furthermore, the IEEE 802.15.6 standard is widely used for enhanced occupational health and environment monitoring applications as it supports high reliability, low power and can manage data rates up to 10Mbps with ease [20]. Other than this, IoT systems make use of three different communication technologies: Bluetooth low energy, near field communication, and radio frequency identification. All these above discussed elements play an essential role in designing and deploying IoT based IAQ monitoring systems. These applications promote enhanced patient care while optimizing resource consumption and the overall cost of healthcare management. Privacy is the prime concern in all IoT applications as patient data must be kept confidential [12]. Therefore, the communication elements, sensing devices, and identification mechanisms must incorporate enhanced methods to ensure the safety, privacy, and reliability of the system.

2.3   IoT Open-Source Platforms and Operating Systems

There are several open-source operating systems (OS) and platforms that extend support to different applications while handling data safety, confidentiality, dissemination, and fusion. This section provides details about some of the most relevant IoT platforms along with their significance in the development of IoT systems.

(1) **SiteWhere**: This open-source IoT platform accelerates the handling, storage, and incorporation of device data. SiteWhere is efficient enough to provide device management, IoT server platform, and third-party integration frameworks. This IoT platform promises enhanced functionalities for automation, monitoring, and analytics in IAQ systems [21].

(2) **Brillo**: This Android-based operating system comes with core services that provide a developer console and developer kit to design and manage IoT applications. It also offers scalability for error reporting, metrics, and OTA updates. Brillo is supported by Intel ×86, ARM, and MISP-based hardware. It is also recognized for providing secure services [22].

(3) **Ubuntu Core:** It is also known as Snappy and it is a development version of Ubuntu that promises enhanced safety and extensibility for the normal ubuntu OS. It delivers management systems for reliable, safe, and transactional updates that are further controlled by Canonical's AppArmor security system [23].

(4) **DeviceHive**: This open-source data platform aims to connect different devices to the cloud and manage device data streams. It also provides customization and creation capabilities for IoT/machine to machine applications while ensuring scalable, secure, and cloud-ready functionalities [24].

(5) **Contiki**: This open-source OS for IoT provides standard IPv4, IPv6, RPL, 6lowpan, and CoAP protocols. It also assists in providing a network simulation environment for IoT applications [25].

(6) **IoTivity**: Here is another open-source framework that maintains the device-to-device communications for IoT applications. The IoTivity projects are managed and sponsored by the Open Connectivity Foundation which is a certification and specification program for addressing IoT related open issues [26].

(7) **DSA**: DSA stands for Distributed Services Architecture which is an open-source platform for joining heterogeneous software and hardware in IoT to provide resilient and scalable decentralized solutions. DSA is made up of DSLink, DSBroker, node API where DSBroker works as a router for outgoing and incoming streams, NodeAPI serves node compatibility and bi-directional control with monitoring ability between connected things. DSLink establishes a connection with DSBroker while acting as a source for different data streams [27].

(8) **Platformio**: This integrated development environment supports cross-platform build functionality. It does not require any external dependency

on OS software and is compatible with 15+ development platforms, 200+ embedded boards, and 10+ frameworks. Other than this, it also offers a built-in serial port monitor, automatic firmware uploading, and configurable build flags/options for IoT applications [28].

(9)   **Netbeast**: This open-source IoT platform aims to connect different IoT platforms that connect various devices and provide agile development opportunities for IoT applications. It works perfectly well with more than 30 unique smart home devices and ten popular brands including Parrot, Google Chromecast, Belkin Wenmo, and Philips Hue [29].

(10) **Kaa**: This multi-purpose middleware platform delivers relevant tools for software development with enhanced features to support IoT applications, it promises decreased cost, time to market, and reduces risks as well. This agonistic hardware solution also extends support to SDK for a diverse range of programming languages including JAVA, C++, and C [30].

(11) **RIOT**: RIOT is a free and open-source OS that works with the majority of open standards to support IoT applications. It ensures enhanced code compatibility for 32-bit, 16-bit, and 8-bit platforms while providing better energy-efficiency, real-time capability. It also provides IPv6, 6LoWPAN routing protocols for lossy and low-power networks [31].

(12) **Calvin-Base**: This open-source platform is designed with a centralized architecture and it extends support to REST API that is highly scalable for implementing a variety of plugins to achieve enhanced interoperability [32].

(13) **ThingsBoard**: An open-source platform for data processing, collection, visualization, and device management. It supports device connectivity via standard IoT protocols such as HTTP, CoAP, and MQTT. Moreover, ThingsBoard extends support to data processing rule chains along with alarm configurations depending upon attribute updates, events, device inactivity, and different user actions [33].

(14) **ARM mbed**: This platform delivers cloud facilities, OS, tools, and designer ecosystems to assist in the development of scalable systems for IoT applications. It also implements safety functionalities including RESTful API, CoAP, and Transport Layer Security for M2M network design [34].

(15) **Cyclon.js**: This JavaScript framework for IoT makes use of Node.js while providing code compatibility among different hardware. It also supports multiple platforms like Raspberry, Intel Edison, Intel Galileo, and Arduino [35].

(16) **ThingSpeak**: ThingSpeak can process HTTP requests and it can process data as well as store information for future use. This open data platform comes with real-time data collection ability and open API. It can also assist users with data geolocation, visualization, plugins and device status messages [36].

(17) **OpenRemote**: OpenRemote has four unique integration tools for manufacturers, distributors, integrators and home-based hobbyists. It supports several protocols and allow users to create and control any smart device [37].

(18) **Nimbits**: It is capable enough to process and store geo stamped or time-stamped data. This public platform can be used as a service online or users

can download it in the form of software to deploy on any Google App Engine. It supports multiple languages for programming including Nimbits.io Java library, HTML, JavaScript and Arduino [38].

(19) **Raspbian**: The developers of this platform are already working on multiple IoT related protocols and sensor networks. Raspbian OS is owned and developed by Raspberry Pi and it offers a highly flexible solution for experimental, and educational needs [39].

(20) **Particle**: Particle is another cloud-based, distributed IoT OS. It can help users to connect thousands of IoT devices, extract and manage data on a real-time basis [40].

IoT platforms are mainly designed to handle specific applications from different domains and they can handle a variety of operations such as device management support, data collection, security, analytics, storage, visualization, and integration [41]. IoT platforms reduce system complexity while increasing the compatibility and scalability of the design. Furthermore, they provide connectivity between application layers and hardware. As there is an extensive range of OS and platforms in the IoT domain, it is not possible to discuss them all in this chapter. Therefore, 20 OS and platforms are discussed above. This analysis shows that the majority of these platforms support IoT device management which plays a crucial role in ever-growing IoT architectures. The main attributes can be MAC addresses, serial numbers, firmware versions, and the location of the device. These attributes can further help in creating groups of devices deployed at specific locations while establishing their communication channel with the other group. Another essential feature related to device management is allowing remote access and maintaining an authorization mechanism.

Existing studies reveal that Contiki, Brillo, Cylon.js, Calvin-Base, and RIOT protocols do not support native security capabilities [5]. However, it is possible to use third-party plugins to achieve security goals. The data integration and collection features are mostly supported by REST and MQTT APIs. Data visualization and analysis are two important features as IoT devices generate a massive amount of data. However, ARM mbed, RIOT, Platformio, Cylon.js, and Ubuntu Core do not offer analytics support. It is relevant to process all the recorded raw and unstructured data to create a structured version of data before applying visualization, pattern analysis, analytics, and charting techniques. As analytics feature is essential to conduct clinical diagnosis and analysis; therefore, these platforms are generally not recommended for IAQ applications. Furthermore, ARM mbed, platformio, RIOT, IoTivity, Cylon.js, Contiki, and Calvin-based platforms lack in terms of storage features.

Kaa platform supports encryption channels, open protocols for data security while providing enhanced data storage, visualization, and analytics capabilities. Therefore, can be recommended for IAQ monitoring applications. This platform also has third-party integrations, is scalable, and supports open IoT protocols including JSON encoding, CoAP, and MQTT. Other than this, it also offers enhanced gateway support and enhanced security over device-level communications via datagram transport layer security or TLS. Kaa also provides flexible application versioning while incorporating well-tested and reliable open-source components.

Several researchers these days are also using ThingSpeak for obtaining field data from IoT environment [42, 43]. It is rated high for interactive user interface and integration to Matlab development platform which makes data analysis and processing much easier. IoT architectures with all such features can be applied for IAQ monitoring applications to implement efficient systems for patient monitoring and elderly support.

2.3.1   Cost-Effective Sensors

IAQ monitoring systems require constant monitoring with calibrated, low-cost and reliable sensors. Several manufacturers have designed IoT based sensors to monitor potential air pollutants such as SO_2, NO_2, PM_{10}, $PM_{2.5}$, VOC, CO_2, and CO [44, 45]. However, the selection of the best sensor often requires in-depth research on essential features, specifications, and availability of the product [44]. Furthermore, the compatibility of selected sensors with the microcontrollers and gateway to design a complete IoT system must be analysed [46]. IAQ levels can be measured using a variety of sensors; some of the widely recommended options for measuring PM concentration are laser sensors [43]. Harmful gaseous pollutants present in the indoor environment can be measured using analog and digital sensors. However, negative and positive temperature coefficient-based sensors are used to measure temperature and humidity in the premises [47].

Several IAQ sensors used in the target environments generate data with standard units such as Celsius and ppm. However, others provide an analog signal which can be later converted into standard units. While making a selection of sensors, researchers also need to think about their calibration and accuracy levels [45]. It is possible to find a wide range of sensors that come in factory calibrated form and can be deployed easily in the target environment; however, these products are usually expensive. On the other side, the sensors that come without calibration can be calibrated before field deployment by researchers [48]. Furthermore, the design of IoT-based sensor networks also requires a detailed analysis of power consumption, reliability, and field performance. It is extremely important to note that environmental monitoring applications require consistent monitoring. Therefore, the sensors must be efficient to provide real-time readings for 24×7 h without facing failures.

2.3.2   End-User Devices

The design and application of IoT based monitoring systems are not limited to measuring field data, they must provide real-time access to the recorded values. The IAQ monitoring systems are desired to provide real-time updates about field data to the building occupants [49]. Consequently, it is essential to interface these systems with some end-user devices. The interfacing is usually done through a gateway unit which connects sensors to the data storage system via a microcontroller. The data storage can be done locally or using cloud-based platforms as well. On the other hand, the end-user devices can be smartphones, tablets, and desktop computers. It is possible to design interactive and user-friendly mobile apps to help users get instant updates about pollutant concentrations. Furthermore, cloud storage systems also make it easier to access data through websites. The login features also make it easier to prevent unauthorized access to data obtained from the indoor environment. Finally, the end-user devices must also get alerts from associated IoT monitoring systems regarding critical threshold levels.

2.4   IoT in IAQ Monitoring: Challenges and Open Issues

IAQ and public health must incorporate advanced IoT mechanisms to ensure enhanced functionality and efficiency of the system in real-time environments. The design and deployment of IoT systems for IAQ monitoring applications require to focus on several aspects such as security measures, architecture, network protocols, scheduling algorithms, and memory management. In this section, we are going to analyze the most important aspects of IoT device OS along with relevant challenges and open issues in detail.

2.4.1   Network Connectivity

As the potential of IoT technology is growing, in the future, it is expected to connect trillions of physical objects and devices to the internet. In this scenario, reliable and stable internet connectivity is the main requirement for IoT applications. The highly efficient and cost-effective IoT systems must support low power, uninterrupted, and internet-enabled communication stack. These channels must be flexible for different configurations to meet the changing demands of the IoT application domain. Furthermore, it is crucial to involve IPv6 protocol in design to maintain the unique identities of devices over large networks. The authors propose a standardized approach with the help of IEEE 802.15.4 standards as it ensures power-efficient and reliable operation. The 6LoWPAN adaptation layer can further enable universal internet connectivity, IETF ROLL routing protocol promises availability and IETF CoAP enables seamless support to internet applications. Most of the authors in the past have used Wi-Fi for communication between IoT based monitoring systems and data storage platforms [50, 51]. However, the main trouble with Wi-Fi is its higher energy consumption requirements. Other potential options in the market with low power requirements are Bluetooth [52, 53] and ZigBee [54, 55]. The main difference between Wi-Fi, Bluetooth and Zigbee lies in terms of its coverage. Wi-Fi is suitable for long-distance transmission (100 m) whereas Bluetooth (<10 m) and Zigbee (10–100 m) have a smaller range [56]. Furthermore, Wi-Fi can support a large number of devices and the frequencies used for this communication are comparatively stable. Other than this, the IoT monitoring systems can be also based on Near Field Communication (NFC) [57, 58] and LoRa (Long Range) [59, 60]. The modern smart monitoring systems are powered by Lora because it can send data up to long distances even while consuming lesser power. In urban environments, it can transmit data up to 2–3 km whereas, in the obstacle-free rural areas, the transmission can reach beyond 5–7 km. Therefore, LoRa is being widely adopted for remote monitoring applications [61, 62].

2.4.2   Heterogeneous Hardware

Monitoring applications require developers to integrate a variety of physical objects and devices over a single network. Therefore, it is critical to find support for heterogeneous hardware elements [63]. The researchers over the years have used different types of IoT nodes such as OpenMote, Arduino, and TelosB [5]. IoT devices also have various microcontroller architectures that may vary from 8-bit to 16-bit and 32-bit designs. Some of the widely used microcontrollers for IoT applications in these ranges are ARM7, TI MSP430, and Intel 8051/8052. These microcontrollers come with variable ROM and RAM sizes while extending support to different communication technologies. Therefore, the selected OS must extend support to the heterogeneous sensor network.

2.4.3   Memory Management

Memory management is another essential aspect of IoT systems. As these networks are likely to generate a massive amount of data, it is significant to find a reliable source for the storage and management of that raw information [17]. However, the IoT devices generally have small memory capacity ranging between KB level of RAM and MB level of Flash. Therefore, it is important to look for advanced memory management to enhance system capabilities. IoT OS must provide a set of cross-layer and optimized libraries that lead to efficient data structures and enhanced IoT functionality. Several researchers in the past have used Raspberry Pi for designing real-time remote monitoring systems because of their higher memory capacity [64, 65]. The Raspberry Pi 2 and Raspberry Pi 3 comes with 1 GB memory, whereas Raspberry Pi 4 is available with 2, 4 and 8 GB RAM version. For comparatively smaller data collection applications, existing researchers also preferred using Arduino Uno which comes with 32 kB of flash memory and 2 kB of SRAM [66, 67]. Furthermore, Arduino Mega comes with 256 kB flash memory, 8 kB SRAM and 4 kB EEPROM [68, 69]. Along with memory capacity, selection of

MCU also depends upon a number of input/output pins required, power requirements and type of processor. On the other hand, numerous Arduino shields with SD card support are available on the market.

2.4.4 Security and Privacy

One of the major challenges in the real-time implementation of IoT systems is the safety weaknesses of M2M communication channels. The developers also need to ensure authorized access to the recorded data so that only the right individuals can process details. Future researchers need to pay more attention to privacy, security, and legal aspects associated with IoT applications [16]. As smart monitoring systems are mostly wireless, the collected data must be encrypted or protected to ensure the desired level of authentication about patient health conditions [12]. It is crucial to develop both, hardware and software level, encryptions to support privacy and safety policies. Authors propose two reliable security schemes to handle this issue: first is that recorded data must be stored at an IoT authentication server and second, all the information must be encrypted using standard procedures [44]. Other than this, it also demands serious efforts to eliminate software attacks and cybercrime-related vulnerabilities. The security protocols and privacy policies must be applied to every layer of the network architecture including the application layer, transport layer, data-link layer, and physical layer [70].

2.4.5 Usability and Interfacing

As IoT systems interact with the users, the architecture is required to be user-friendly while offering easy interfacing with a variety of end devices. All commercial, consumer, industrial, healthcare, and IAQ monitoring applications of IoT must be aligned with enhanced user experience to lead to rapid growth. Hardware and software development phases must focus on cross-platform operations with enhanced system compatibility. The millennials also demand personalization features in the IoT applications, especially in the behavioral domains [71]. Figure 2.2. provides quick insights into potential challenges and open issues for IoT in the environmental monitoring field.

Despite several efforts made by the existing researchers to enhance the capabilities of IoT systems, particularly in the IAQ monitoring domain, several open issues persist; Self-management, user privacy, identity management, network protocol, cryptographic mechanisms, interoperability, and scalability are some common examples [41, 71]. Several studies in the past have also put light on the importance of solving these problems on a priority basis to improve the overall quality of the IoT based IAQ monitoring systems [15, 70, 71].

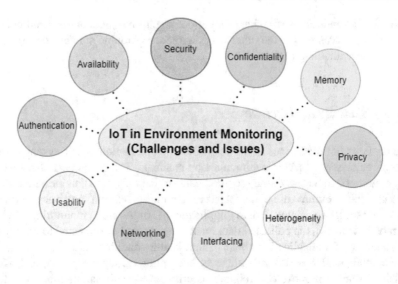

Fig. 2.2 IoT for environment monitoring: challenges and issues

2.5   Discussion and Future Directions

IoT is recognized as a relevant technology for IAQ monitoring and public health management. However, several challenges and open issues still exist in this technology area that is required to be addressed in near future. Sincere research efforts are required to address the user interface, system usability, ergonomics, data structure, data access, security, privacy, and several social as well as ethical issues. IoT architectures are responsible for transferring a massive amount of data over the internet, without even guaranteeing data privacy. Therefore, several social, ethical, and legal questions are raised regarding the transparency, security, and privacy of IoT systems. This particular issue plays a crucial role in IAQ monitoring applications as patient data should not be accessed by unauthorized sources. Pervasive monitoring applications focus on collecting sensitive data related to their daily route and overall well-being. These smart monitoring systems must incorporate advanced security mechanisms to guarantee complete reliability and safety of data. The policymakers and government authorities also need to develop strong regulations and legislations to guarantee user rights, especially in the IAQ domain. While adopting new policies for the design of IoT systems, developers need to focus on ethics as a potential actor to deliver trustworthy and safe infrastructures. The data for IAQ monitoring systems must be protected and implemented by using encryption schemes, appropriate cryptographic, key management, and security measures to avoid privacy leakages. Furthermore, IoT devices in the smart home monitoring sector must be verified on regular basis to avoid vulnerabilities and security attacks.

The real-time monitoring systems deal with an extensive amount of information. Therefore, extra efforts must be made to deal with the diversity and complexity of data. Furthermore, it is crucial to follow uniform standards for heterogeneous data management. The diverse data sources also need a unified programming interface to handle multiple data analysis modules. Effective and efficient cooperation between digital and physical domains must be established to ensure the reliable working of IAQ monitoring systems. The advanced technologies such as 5G also open new opportunities for growth and wide acceptance of IoT based IAQ systems as it supports real-time applications involving video on demand for enhanced patient care.

The main goal of smart IoT developments is to present intelligent and effective systems for monitoring health and well-being. The developers also need to be careful while making a selection for the configuration and design settings for effective data acquisition. A wide range of sensors including wearable sensors, gas sensors, and particle pollutant sensors are commonly used for the design of IAQ monitoring systems. These sensors can be deployed in residential apartments, commercial buildings, gyms, schools, and hospitals. Smartwatches, tablets, and smartphones can be used to access the data collected through sensors and these expert systems further use built-in sensors such as a microphone, camera, and GPS to transfer information to the externally interfaced devices. Moreover, assistive robots are also used in combination with ambient sensors for enhanced monitoring and activity recognition.

The researchers also need to pay more attention to the use of secure data transmission techniques to prevent threats to privacy [12]. There are some widely used methods for wireless communication and data sharing: ZigBee and Bluetooth are generally preferred for device-to-device communication; mobile networks and Wi-Fi technologies such as 3G and 4G are used for providing internet access [70, 72]. The new age IoT systems provide mobile access to the recorded data via web interfaces. However, IAQ systems need to incorporate security methods to avoid unauthorized access.

The ongoing advancements in IAQ monitoring systems make them truly promising with real-time environment monitoring capabilities. These systems can be widely used for real-time notifications, pollutant threshold level tracking, and sending alerts to the medical teams to manage critical conditions. IoT based real-time monitoring systems can be also used to monitor the living environment to enhance occupant health and well-being. Furthermore, it is important to design cost-effective, reliable, and efficient IAQ solutions with the support of open-source technologies to improve occupational health. There is also a need to work on several limitations such as validation, energy consumption, and calibration to address the real-time scenarios. Moreover, the future monitoring systems also need to send instant notifications to the users to avoid critical consequences associated with the monitoring environment.

2.6 Conclusion

IoT technology offers advanced architectures, methods, and solutions for the development of effective IAQ monitoring systems. It offers a reliable opportunity to improve the overall quality of occupational health with personalized environment monitoring. The open-source platforms and feature-rich OS make it easier to improve the availability, security, and quality of smart monitoring systems. Moreover, the variety of sensor modules can further enhance the efficiency and evolution of IAQ monitoring systems while enabling highly accurate device-level monitoring. Cost-effective platforms and open-source tools can also assist in real-time monitoring with smartphone or smartwatch compatibility. These IoT systems can also provide mobile notifications and instant alerts to improve the user's health and well-being.

Despite the numerous advantages of IoT-based IAQ monitoring systems, several challenges and open-issues also exist. Future researchers need to work on the scalability, performance, mobility, reliability, availability, and interoperability of systems to ensure enhanced support to the end-users. IoT based IAQ systems must support ventilation arrangements while complementing pollutant concentration supervision. This chapter presented essential aspects relevant to IoT in the IAQ domain while discussing potential open issues, challenges along with details about potential OS and open-source platforms. It is expected that this overview could be useful for IT professionals, students, and engineers while extending support to future research in the indoor environment and occupational health sector.

References

1. G. Marques, N. Miranda, A. Kumar Bhoi, B. Garcia-Zapirain, S. Hamrioui, I. de la Torre Díez, Internet of things and enhanced living environments: measuring and mapping air quality using cyber-physical systems and mobile computing technologies. Sensors **20**(3) (2020), Art. no. 3. https://doi.org/10.3390/s20030720
2. P. Koleva, K. Tonchev, G. Balabanov, A. Manolova, V. Poulkov, Challenges in designing and implementation of an effective ambient assisted living system, in *2015 12th International Conference on Telecommunication in Modern Satellite, Cable and Broadcasting Services (TELSIKS)*, Oct 2015, pp. 305–308. https://doi.org/10.1109/TELSKS.2015.7357793
3. Y. Yin, Y. Zeng, X. Chen, Y. Fan, The internet of things in healthcare: an overview. J. Ind. Inf. Integr. **1**, 3–13 (2016). https://doi.org/10.1016/j.jii.2016.03.004
4. L. Atzori, A. Iera, G. Morabito, The internet of things: a survey. Comput. Netw. **54**(15), 2787–2805 (2010). https://doi.org/10.1016/j.comnet.2010.05.010
5. G. Marques, R. Pitarma, N.M. Garcia, N. Pombo, Internet of things architectures, technologies, applications, challenges, and future directions for enhanced living environments and healthcare systems: a review. Electronics **8**(10), 1081 (2019). https://doi.org/10.3390/electronics8101081
6. M. Memon, S.R. Wagner, C.F. Pedersen, F.H.A. Beevi, F.O. Hansen, Ambient assisted living healthcare frameworks, platforms, standards, and quality attributes. Sensors **14**(3) (2014), Art. no. 3. https://doi.org/10.3390/s140304312
7. T. Tsukiyama, In-home health monitoring system for solitary elderly. Proc. Comput. Sci. **63**, 229–235 (2015). https://doi.org/10.1016/j.procs.2015.08.338
8. J. Lee, Y. Chuah, K.T.H. Chieng, Smart elderly home monitoring system with an android phone. Int. J. Smart Home **7**(3), 17–32 (2013)
9. I. Azimi, A.M. Rahmani, P. Liljeberg, H. Tenhunen, Internet of things for remote elderly monitoring: a study from user-centered perspective. J. Ambient Intell. Human Comput. **8**(2), 273–289 (2017). https://doi.org/10.1007/s12652-016-0387-y
10. P.P. Ray, Home health hub internet of things (H3IoT): an architectural framework for monitoring health of elderly people, in *2014 International Conference on Science Engineering and*

Management Research (*ICSEMR*), Nov 2014, pp. 1–3. https://doi.org/10.1109/ICSEMR.2014. 7043542

11. A. Dohr, R. Modre-Opsrian, M. Drobics, D. Hayn, G. Schreier, The internet of things for ambient assisted living, in *2010 Seventh International Conference on Information Technology: New Generations*, Apr 2010, pp. 804–809. https://doi.org/10.1109/ITNG.2010.104

12. S. Al-Janabi, I. Al-Shourbaji, M. Shojafar, S. Shamshirband, Survey of main challenges (security and privacy) in wireless body area networks for healthcare applications. Egypt. Inf. J. **18**(2), 113–122 (2017). https://doi.org/10.1016/j.eij.2016.11.001

13. J. Saini, M. Dutta, G. Marques, A comprehensive review on indoor air quality monitoring systems for enhanced public health. Sustain. Environ. Res. **30**(1), 6 (2020). https://doi.org/10. 1186/s42834-020-0047-y

14. D. Niyato, E. Hossain, S. Camorlinga, Remote patient monitoring service using heterogeneous wireless access networks: architecture and optimization. IEEE J. Sel. Areas Commun. **27**(4), 412–423 (2009). https://doi.org/10.1109/JSAC.2009.090506

15. K. Sha, W. Wei, T. Andrew Yang, Z. Wang, W. Shi, On security challenges and open issues in internet of things. Future Gen. Comput. Syst. **83**, 326–337 (2018). https://doi.org/10.1016/ j.future.2018.01.059

16. S.B. Baker, W. Xiang, I. Atkinson, Internet of things for smart healthcare: technologies, challenges, and opportunities. IEEE Access **5**, 26521–26544 (2017). https://doi.org/10.1109/ ACCESS.2017.2775180

17. D. Chen, G. Chang, L. Jin, X. Ren, J. Li, F. Li, A novel secure architecture for the internet of things, in *2011 Fifth International Conference on Genetic and Evolutionary Computing*, Aug 2011, pp. 311–314. https://doi.org/10.1109/ICGEC.2011.77

18. A.J. Jara, M.A. Zamora, A. Skarmeta, Glowbal IP: an adaptive and transparent IPv6 integration in the internet of things. Mob. Inf. Syst. **8**(3), 177–197 (2012). https://doi.org/10.1155/2012/ 819250

19. M. Goyal, W. Xie, H. Hosseini, IEEE 802.15.4 modifications and their impact. Mob. Inf. Syst. **7**(1), 69–92 (2011). https://doi.org/10.3233/MIS-2011-0111

20. S. Saleem, S. Ullah, K.S. Kwak, A study of IEEE 802.15.4 security framework for wireless body area networks. Sensors **11**(2) (2011), Art. no. 2. https://doi.org/10.3390/s110201383

21. SiteWhere Open Source Internet of Things Platform. https://sitewhere.io/en/. Accessed 10 Jan 2021

22. Digital Transformation Consulting I Customer Experience Transformation, in *Brillio*. https:// www.brillio.com/. Accessed 10 Jan 2021

23. Ubuntu Core, *Ubuntu*. https://ubuntu.com/core. Accessed 10 Jan 2021

24. DeviceHive—Open Source IoT Data Platform with the wide range of integration options. https://devicehive.com/. Accessed 10 Jan 2021

25. Contiki Travel Tours I Adventure Holidays for 18–35 year olds, in *Contiki*. https://www.con tiki.com/ap/en. Accessed 10 Jan 2021

26. Home I IoTivity. https://iotivity.org/. Accessed 10 Jan 2021

27. Open Source IoT Platform and Toolkit I DSA—Home. http://iot-dsa.org/. Accessed 10 Jan 2021

28. PlatformIO, PlatformIO is a professional collaborative platform for embedded development, in *PlatformIO*. https://platformio.org. Accessed 10 Jan 2021

29. Luis Pinto Feed 404up, Netbeast: a cross-platform tool for the Internet of Things, in *Opensource.com*, 06 May 2016. https://opensource.com/life/16/5/netbeast (accessed Jan. 10, 2021).

30. Enterprise IoT Platform with Free Plan I Kaa, in *Kaa IoT Platform*. https://www.kaaproject.org. Accessed 10 Jan 2021

31. H. Will, K. Schleiser, J. Schiller, A real-time kernel for wireless sensor networks employed in rescue scenarios, in *2009 IEEE 34th Conference on Local Computer Networks*, Zurich, Switzerland, Oct 2009, pp. 834–841. https://doi.org/10.1109/LCN.2009.5355049

32. *EricssonResearch/calvin-base* (Ericsson Research, 2021)

33. Thingsboard, Thingsboard—open-source IoT platform, in *ThingsBoard*. https://thingsboard.io/. Accessed 10 Jan 2021
34. *Fast and Effective Embedded Systems Design* (Elsevier, 2017)
35. Cylon.js—JavaScript framework for robotics, physical computing, and the Internet of Things using Node.js. https://cylonjs.com/. Accessed 10 Jan 2021
36. IoT Analytics—ThingSpeak Internet of Things. https://thingspeak.com/. Accessed 20 Jan 2021
37. OpenRemote | The 100% open source IoT platform, in *OpenRemote*. https://openremote.io/. Accessed 20 Jan 2021
38. Business profile for nimbits.com provided by network solutions. http://www.nimbits.com/. Accessed 20 Jan 2021
39. FrontPage—Raspbian. https://www.raspbian.org/. Accessed 20 Jan 2021
40. Particle Company News and Updates, in *Particle*. https://www.particle.io/. Accessed 20 Jan 2021
41. P.P. Ray, A survey of IoT cloud platforms. Future Comput. Inf. J. **1**(1), 35–46 (2016). https://doi.org/10.1016/j.fcij.2017.02.001
42. S. Pasha, Thingspeak based sensing and monitoring system for IoT with Matlab analysis. Int. J. New Technol. Res. (IJNTR) **2**(6), 5 (2016)
43. J. Saini, M. Dutta, G. Marques, Indoor air quality monitoring with IoT: predicting PM_{10} for enhanced decision support, in *2020 International Conference on Decision Aid Sciences and Application (DASA)*, Sakheer, Bahrain, Nov 2020, pp. 504–508. https://doi.org/10.1109/DASA51403.2020.9317054
44. J. Saini, M. Dutta, G. Marques, Sensors for indoor air quality monitoring and assessment through internet of things: a systematic review. Environ. Monit. Assess. **193**(2), 66 (2021). https://doi.org/10.1007/s10661-020-08781-6
45. N. Afshar-Mohajer et al., Evaluation of low-cost electro-chemical sensors for environmental monitoring of ozone, nitrogen dioxide, and carbon monoxide. J. Occup. Environ. Hyg. **15**(2), 87–98 (2018). https://doi.org/10.1080/15459624.2017.1388918
46. G. Marques, J. Saini, M. Dutta, P.K. Singh, W.-C. Hong, Indoor air quality monitoring systems for enhanced living environments: a review toward sustainable smart cities. Sustainability **12**(10), 4024 (2020). https://doi.org/10.3390/su12104024
47. J. Saini, M. Dutta, G. Marques, Internet of things based environment monitoring and PM_{10} prediction for smart home, in *2020 International Conference on Innovation and Intelligence for Informatics, Computing and Technologies (3ICT)*, Sakheer, Bahrain, Dec 2020, pp. 1–5. https://doi.org/10.1109/3ICT51146.2020.9311996
48. ISO and IEC, ISO/IEC 17025:2017 general requirements for the competence of testing and calibration laboratories, Nov 2017
49. S. Abraham, X. Li, A cost-effective wireless sensor network system for indoor air quality monitoring applications. Proc. Comput. Sci. **34**, 165–171 (2014). https://doi.org/10.1016/j.procs.2014.07.090
50. S. Sadu, R.R. Arabelli, IOT based monitoring and control system for appliances, p. 8
51. M. Bassoli, V. Bianchi, I.D. Munari, P. Ciampolini, An IoT approach for an AAL Wi-Fi-based monitoring system. IEEE Trans. Instrum. Meas. **66**(12), 3200–3209 (2017). https://doi.org/10.1109/TIM.2017.2753458
52. L. Fourati, S. Said, *Remote Health Monitoring Systems Based on Bluetooth Low Energy (BLE) Communication Systems* (2020), pp. 41–54
53. R.N. Gore, H. Kour, M. Gandhi, D. Tandur, A. Varghese, Bluetooth based sensor monitoring in industrial IoT plants, in *2019 International Conference on Data Science and Communication (IconDSC)*, Mar 2019, pp. 1–6. https://doi.org/10.1109/IconDSC.2019.8816906
54. D. Chen, M. Wang, A home security Zigbee network for remote monitoring application, in *2006 IET International Conference on Wireless, Mobile and Multimedia Networks*, Nov 2006, pp. 1–4. https://doi.org/10.1049/cp:20061246
55. S. Hwang, D. Yu, Remote monitoring and controlling system based on ZigBee networks. J. Softw. Eng. Appl. **6**, 35–42 (2012)

56. K. Pothuganti, A. Chitneni, *A Comparative Study of Wireless Protocols: Bluetooth, UWB, ZigBee, and Wi-Fi*, vol. 4, Sept 2014, pp. 655–662

57. S. Yadav, V. Chappar, S. Datir, P. Jagtap, A pervasive and personalized smart healthcare system using IOT. Int. J. Adv. Sci. Eng. Technol. (IJASEAT) **5**(3), 6

58. S.S. Shaikh, H.R. Hossein, NFC and IoT based smart m-healthcare patient monitoring system. Biomet. Bioinform. **8**(8) (2016), Art. no. 8

59. G. Pătru, D. Trancă, C. Costea, D. Rosner, R. Rughiniş, LoRA based, low power remote monitoring and control solution for Industry 4.0 factories and facilities, in *2019 18th RoEduNet Conference: Networking in Education and Research (RoEduNet)*, Oct 2019, pp. 1–6. https://doi.org/10.1109/ROEDUNET.2019.8909499

60. C. Choi, J. Jeong, I. Lee, W. Park, LoRa based renewable energy monitoring system with open IoT platform, in *2018 International Conference on Electronics, Information, and Communication (ICEIC)*, Jan 2018, pp. 1–2. https://doi.org/10.23919/ELINFOCOM.2018.8330550

61. A.F. Rachmani, F.Y. Zulkifli, Design of IoT monitoring system based on LoRa technology for starfruit plantation, in *TENCON 2018–2018 IEEE Region 10 Conference*, Oct 2018, pp. 1241–1245. https://doi.org/10.1109/TENCON.2018.8650052

62. N. Misran, M.S. Islam, G.K. Beng, N. Amin, M.T. Islam, IoT based health monitoring system with LoRa communication technology, in *2019 International Conference on Electrical Engineering and Informatics (ICEEI)*, Jul 2019, pp. 514–517. https://doi.org/10.1109/ICEEI47359.2019.8988869

63. K. Xu, Y. Qu, K. Yang, A tutorial on the internet of things: from a heterogeneous network integration perspective. IEEE Netw. **30**(2), 102–108 (2016). https://doi.org/10.1109/MNET.2016.7437031

64. D.E.N. Ganesh, Health monitoring system using raspberry Pi and IOT. Orient. J. Comput. Sci. Technol. **12**(1), 08–13

65. M.D. Mudaliar, N. Sivakumar, IoT based real time energy monitoring system using Raspberry Pi. Internet Things **12**, 100292 (2020). https://doi.org/10.1016/j.iot.2020.100292

66. S. Jayapradha, P.M.D.R. Vincent, An IOT based human healthcare system using Arduino uno board, in *2017 International Conference on Intelligent Computing, Instrumentation and Control Technologies (ICICICT)*, July 2017, pp. 880–885. https://doi.org/10.1109/ICICICT1.2017.8342681

67. A. Nduka, J. Samual, S. Elango, S. Divakaran, U. Umar, R. Senthil Prabha, Internet of things based remote health monitoring system using Arduino, in *2019 Third International Conference on I-SMAC (IoT in Social, Mobile, Analytics and Cloud) (I-SMAC)*, Dec 2019, pp. 572–576. https://doi.org/10.1109/I-SMAC47947.2019.9032438

68. L. Barik, IoT based temperature and humidity controlling using Arduino and Raspberry Pi. IJACSA **10**(9) (2019). https://doi.org/10.14569/IJACSA.2019.0100966

69. Aditya, M. Sharma, S.C. Gupta, An internet of things based smart surveillance and monitoring system using Arduino, in *2018 International Conference on Advances in Computing and Communication Engineering (ICACCE)*, June 2018, pp. 428–433. https://doi.org/10.1109/ICACCE.2018.8441725

70. H.F. Atlam, R.J. Walters, G.B. Wills, Internet of things: state-of-the-art, challenges, applications, and open issues. IJICR **9**(3), 928–938 (2018). https://doi.org/10.20533/ijicr.2042.4655.2018.0112

71. K. Patel, S. Patel, P. Scholar, C. Salazar, *Internet of Things-IOT: Definition, Characteristics, Architecture, Enabling Technologies, Application & Future Challenges* (2016)

72. N.V.R. Kumar, C. Bhuvana, S. Anushya, Comparison of ZigBee and Bluetooth wireless technologies-survey, in *2017 International Conference on Information Communication and Embedded Systems (ICICES)*, Feb 2017, pp. 1–4. https://doi.org/10.1109/ICICES.2017.8070716

Chapter 3
Indoor Air Quality and Internet of Things: The State of the Art

3.1 Introduction

Indoor air quality (IAQ) has been a major concern for public health and well-being. As human beings spend most of their routine time indoors, maintaining good air quality is vital for an improved lifestyle. Studies reveal that indoor air pollutants can usually become two to five times dangerous, occasionally up to 100 times harmful, as compared to the outdoor air pollutants [1]. The impact of these variations is more dangerous on people who are already affected by certain respiratory health problems, infants, disabled people, and the elderly population [2]. Literature provides evidence on the strong association of IAQ with a wide range of health problems such as chronic respiratory disease, poor lung functionality, heart disease, development disorders, and damage to the nervous system, brain, liver, and kidneys [3–5]. Several factors are contributing to decay in IAQ levels including primary and secondary pollutants such as volatile organic compounds (VOC), particulate matter (PM), biological particles, and gases including carbon monoxide, ozone, radon, nitrogen oxide, and sulfur dioxide [6]. Therefore, it is essential to monitor these indoor air pollutants on a real-time basis to inform building occupants about critical threshold levels. These measurements and timely alerts can help end-users to make planned decisions regarding ventilation to ensure a safe level of pollutants within the premises [7].

Evidence reveals that indoor air pollution (IAP) affects people's productivity levels and working performance [8]. It leads to the estimated loss of \$20–\$200 billion every year due to approximately a 5% decay in workplace productivity levels [9]. The impact of decaying pollutant concentration can be truly damaging for the people who spend 80–90% of their routine time indoors. This is because the indoor spaces promote the fast build-up of harmful pollutants as compared to the open spaces [2]. It is observed that almost one half of the global population, especially 95% of people in developing countries rely on solid fuels such as coal and biomass for handling their routine heating and cooking needs. In India, more than 0.2 billion people use biomass fuel for cooking where the main preference is given to firewood, cow dung

© The Author(s), under exclusive license to Springer Nature Switzerland AG 2021
J. Saini et al., *Internet of Things for Indoor Air Quality Monitoring*,
SpringerBriefs in Computational Intelligence,
https://doi.org/10.1007/978-3-030-82216-3_3

cake, liquid petroleum gas, biogas, and kerosene [10]. The improper and incomplete combustion of biomass fuels in the conventional stoves leads to excessive generation of harmful pollutants such as PM, CO_2, NO_x, and toxic organic compounds. These conditions become more critical in homes that have poor ventilation arrangements and can work as a potential cause of chronic health problems [11]. The impact of IAP is not restricted to rural homes. The researchers and scientific community found several pieces of evidence regarding decaying levels of pollutant concentrations in the modern living spaces in urban areas [12–15]. The main sources behind poor air quality in such areas are heating, ventilation, and air conditioning (HVAC) systems, building materials, human activities and repeated use of chemical-rich products. The results of IAP are reported in terms of 2 million premature deaths per year out of which 54% people die from chronic obstructive pulmonary disease (COPD), 44% due to pneumonia and 2% due to lung cancer [10]. To control the negative impact of IAP on building occupants, it is significant to harness the efficiencies of the latest technologies. Several researchers in the past have proposed design and development of highly-efficient, real-time monitoring systems that can provide instant updates about threatening IAP levels to the building occupants [16].

There are mainly two potential methods that support the development of IAQ monitoring systems for residential and commercial spaces: the IoT and wireless sensor networks [17]. As the IoT-based architectures have proven their edge with the ability to handle and transfer bulk data through the internet, they are gaining more popularity for smart building applications. The combination of IoT with modern age communication technologies and hardware solutions support reliable monitoring of IAQ for enhanced environmental health and public well-being [18]. The monitoring systems include a set of sensors to measure desired IAQ parameters, a microcontroller to collect data from sensor units, gateways, and communication systems. The general architecture of IoT monitoring system is given in Fig. 3.1. The data obtained from monitoring systems can be further stored on the cloud for easy access in the future. It can help building occupants to take a relevant decision about improving air quality in the building premises.

3.2 IAQ Sensors and Parameters

The monitoring of pollutants in indoor environments requires an adequate selection of sensors. The building premises can have a variety of pollutants and the concentrations can cross critical threshold levels depending upon the human activities, building materials, and heating and cooking arrangements [2]. The existing studies in the literature show that IAQ is greatly affected by several thermal comfort parameters, especially temperature and humidity [19, 20]. The most commonly analyzed IAQ parameters are PM_{10}, $PM_{2.5}$, CO_2, CO, VOC, and NO_x [16, 21–25]. To measure these parameters, researchers have used a variety of sensors including laser sensors, digital sensors, and analog sensors [26, 27]. Most of the researchers over the years preferred using DHT11 and DHT22 sensors for measuring temperature and humidity thermal comfort parameters [28–33]. However, SHT21 is another commonly used

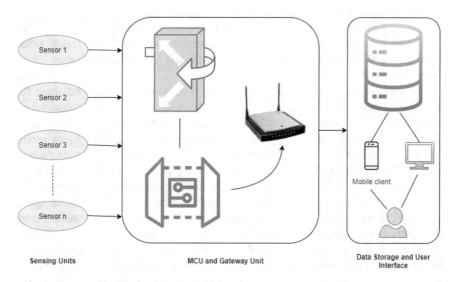

Sensor 1

Sensor 2

Sensor 3

Sensor n

Mobile client

Sensing Units

MCU and Gateway Unit

Data Storage and User Interface

Fig. 3.1 General architecture of IoT based IAQ monitoring system

sensor for measuring thermal variations in the indoor environment [32, 34]. The measurement range of the DHT11 sensor usually lies between 0 and 50 °C whereas the humidity range varies from 20 to 90% [35–37]. Similarly, for DHT22, the temperature measurement varies between −40 and 80 °C and humidity ranges from 0 to 100% [28, 33]. Besides, the manufacturers have specified temperature measurement for SHT21 from −40 to 125 °C and humidity can fall between 0 and 89% [34]. Usually, DHT22 is considered as a more reliable choice for measuring thermal comfort parameters due to its higher accuracy (±0.5 °C; <±2–5% RH) [25]. These sensors can be ordered online from some popular marketplaces such as eBay, Amazon, and Flipkart [25]. All these widely used thermal comfort sensors are available in factory calibrated form. However, the cost may vary as per market conditions. Other studies in the literature mention use of BME280 and SHT10 sensor for measuring thermal comfort parameters [32, 38–40]. The relevant advantage of using BME280 is that it can also measure air pressure along with temperature and humidity.

To measure potential indoor air pollutants, literature provides evidence for using single parameter sensors and multi-parameter sensors as well [16, 25]. PM concentrations are reported to leave a critical impact on human health and well-being. Particularly the elderly, children, women, and disabled people who spend more time indoors. It is possible to find PM in the indoor environment in three different forms including particles with diameters less than 10 μm (PM_{10}), particles with less than 2.5 μm diameter ($PM_{2.5}$), and particles with a diameter below 0.1 μm ($PM_{0.1}$) [41, 42]. Existing studies reveal that PM_{10} and $PM_{2.5}$ are the most preferred measuring parameters for indoor environments [43–45, p. 10]. Several researchers recommend using SHINYEI PPD42NS [30, 46], Sharp GP2Y1010AU0F [31, 35,

47, 48] and PMS5003 Dust Sensor [48–50] to measure PM levels. The trade-off for selection usually goes in terms of cost, measurement range, and accuracy. Other than this, several researchers also preferred using the SDS011 NOVA laser sensor for measuring PM levels [28, 51].

Gaseous pollutants are another considerable challenge for human health in the building premises [52]. The MQ series sensors are recommended by most researchers for making cost-effective measurements of gases in the indoor environment [25]. However, MQ135 is the widely preferred multi-gas sensor for IAQ measurements [36, 37, 47, 53]. It is capable enough to measure six different parameters such as CO_2, C_6H_6, C_2H_5OH, NO_X, NH_3, and smoke. Although this sensor is a reliable choice in terms of cost, operating conditions, and response time; it requires efforts for calibration before field deployment. MQ6 sensor can measure C_2H_2OH, C_3H_8, C_4H_{10}, LPG, and smoke [54]. Other than this, the available options from the same manufacturer are MQ9, MQ5, MQ6, MQ7, MQ2, MQ3, and MQ4 [55–57]. None of these sensors come in factory calibrated form. Instead, they require users to make gas chamber-based calibration before installation. However, the cost for most of these sensors is usually less than \$10 on most of the online marketplaces. For multi-gas measurements, researchers can also prefer MiSC series sensors (6841 and 4514) or go ahead with the Figaro series (TGS2610 and TGS2620) [54, 58]. DFRobot Gravity BME680 sensor is widely recommended for measuring VOC [59, 60]. The main advantage of using this sensor is that it can eliminate the need for investing in additional thermal comfort sensors as it is capable enough to measure atmospheric pressure, humidity, and temperature. Table 3.1 gives quick highlights about commonly used sensors, their measuring parameters and manufacturer specifications.

While making a selection for the sensors to measure IAQ levels, future researchers also need to focus on the overall system performance and error rate for real-time measurements. To maintain reliable performance and adequate readings, the first most requirement is to work on calibration requirements. Even if the sensors are available in factory calibrated form, the authors recommend using reliability tests before field deployment to ensure the desired accuracy in readings [64]. For long-term monitoring applications, it is also recommended to check the frequent calibration requirements of these sensors. Several sensors ensure reliable performance for remote monitoring as they do not consume extensive energy. However, maintaining a sustainable source of power is a must for most IAQ applications.

3.3   Microcontrollers and Gateways

The development of IAQ monitoring systems using IoT architecture require the selection of reliable microcontrollers. Through the study of existing literature, it is found that Raspberry Pi and Arduino has been the most preferred choice for designing gateway and slave MCUs [28, 31, 34, 39, 47, 63, 65–68]. Other than this, many researchers also preferred using ESP8266 module for the gateway development [29,

Table 3.1 Widely used IAQ sensors and their manufacturer specifications

Sensor name	Target parameters	Operating conditions	Measurement range	Calibration requirements	References
DHT11	Temp, Hum	Temp = 0–50 °C; Hum = 20–90%	Temp = 0–50 °C; Hum = 20–90% RH	Pre-calibrated	[30, 31, 35–37, 51]
DHT22	Temp, Hum	Temp = −40 to + 80 °C; Hum = 0–100% RH	Temp = −40 to +80 °C; Hum = 0–100% RH	Pre-calibrated	[28, 29, 32, 33]
SHT21	Temp, Hum	Temp = −40 to 125 °C; Hum = 0–100% RH	Temp = −40 to 125 °C; Hum = 0–80% RH	Pre-calibrated	[32, 34, 59]
SHT10	Temp, Hum	Temp = −40 to 123.8 °C; Hum = 0–100% RH	Temp = −40 to 123.8 °C; Hum = 0–100% RH	Pre-calibrated	[38]
BME280	Temp, Hum, Pressure	Temp = −40 to 85 °C; Hum = 0–100% RH	Temp = 0–60 °C; Hum = 0–100%	Pre-calibrated	[32, 39, 40]
SHINYEI PPD42NS	$PM_{2.5}$	Temp = 0–45 °C; Hum = 0–95% RH	0–28,000 pcs/l	Require calibration	[30, 46]
Sharp GP2Y1010AU0F	PM_{10}, $PM_{2.5}$	Temp = −10 to 65 °C	0–600 $\mu g/m^3$	Require calibration	[31, 35, 47, 56, 58]
PMS5003 Dust Sensor	$PM_{2.5}$	Temp = −10 to + 60 °C; Hum = 0–90% RH	0–500 $\mu g/m^3$	Pre-calibrated	[48–50]
SDS011	PM_{10}, $PM_{2.5}$	Temp = −10 to + 50 °C; Hum = 0–70% RH	0.0–999.9 μ g/m^3	Require calibration	[28, 51, 61]

(continued)

Table 3.1 (continued)

Sensor name	Target parameters	Operating conditions	Measurement range	Calibration requirements	References
MQ135	NH_3, No_x, C_2H_5OH, C_6H_6, CO_2, Smoke	Temp = 20 ± 2 °C; Hum = 65 ± 5% RH	10–300 ppm (NH_3), 10–1000 ppm (C_6H_6), 10–300 ppm (C_2H_5OH)	Require calibration	[22, 31, 35, 37, 47, 48, 53, 56]
MQ9	CO, combustible gas	20 ± 2 °C; 65 ± 5% RH	10–1000 ppm CO; 100–10,000 ppm combustible gas	Require calibration	[56]
MQ7	CO	Temp = −20 ± 2 °C; Hum = 65% ± 5% RH	20–2000 ppm	Require calibration	[36, 38, 48, 53, 56, 57]
MQ6	LPG, C_4H_{10}, C_3H_8, C_2H_5OH, Smoke	Temp = 20 ± 2 °C; Hum = 65 ± 5% RH	200–10,000 ppm	Require calibration	[35, 62]
MQ5	LPG, natural gas, town gas	Temp = 20 ± 2 °C; Hum = 65 ± 5% RH	200–10,000 ppm	Require calibration	[55]
MQ4	CH_4	Temp = 20 ± 2 °C; Hum = 65 ± 5% RH	200–10,000 ppm	Require calibration	[35]
MQ3	C_2H_5OH, C_6H_6	Temp = 20 ± 2 °C; Hum = 65 ± 5% RH	0.04–4 mg/L	Require calibration	[57]
MQ2	SnO_2 (combustible gas) and smoke	Temp = 20 ± 2 °C; Hum = 65 ± 5% RH	300–10,000 ppm	Require calibration	[57]

(continued)

Table 3.1 (continued)

Sensor name	Target parameters	Operating conditions	Measurement range	Calibration requirements	References
MiSC 6841	CO, NO_2, C_2H_5OH, H_2, NH_3, CH_4, C_3H_8, C_4H_{10}	Temp = -30 to $+85$ °C; Hum = 5–95% RH	1–1000 ppm, 0.05–10 ppm, 10–500 ppm, 1–1000 ppm, 1–500 ppm, >1000 ppm, >1000 ppm, >1000 ppm, Resp	Require calibration	[54]
MiSC 4514	CO, NO_2, C_2H_5OH, H_2, NH_3, CH_4,	Temp = -30 to 85 °C; Hum = 5–95% RH	1–1000 ppm, 0.05–10 ppm, 10–500 ppm, 1–1000 ppm, 1–500 ppm, >1000 ppm Resp.	Require calibration	[58]
Figaro TGS2610	LPG, C_4H_{10}	Temp = 20 ± 2 °C; Hum = $65 \pm 5\%$ RH	1–25% LEL	Require calibration	[63]
Figaro TGS2620	Solvent Gases, C_2H_5OH	Temp = 20 ± 2 °C; Hum = $65 \pm 5\%$ RH	50–5000 ppm EtOH	Require calibration	[63]
DFTRobot Gravity BM680	VOC, Temp, Hum, atmospheric pressure	Temp = -40 to $+85$ °C; Hum = 0–100% RH	Temp = -40 to $+85$ °C; Hum = 0–100% RH; Atm Pres = 300–1100 hPa; IAQ = 0–500	Require calibration	[54, 59, 60]

32, 36, 48, 54, 56]. If the IAQ monitoring projects have higher budgets, researchers can also go ahead with Waspmote as an MCU [39, 69]. ESP32 is another preferred choice for the gateway unit [70]. However, researchers in the past preferred ESP8266 more due to its easy interfacing and efficient wireless connectivity [22]. There are multiple versions of Arduinos available in the market and each one of them comes with unique features, specifications, and prices. Arduino Uno is generally the most commonly used microcontroller [71]. It comes with an ATmega328 processor with extended support to 32 KB program memory, 2 KB RAM, 1 KB of EEPROM. This microcontroller has 14 digital input–output pins along with 6 analog inputs. It can be connected on both 3.3 and 5 V power rails for field operations [72, 73]. Arduino Nano on the other side is a smaller version of Arduino Uno which is suitable for tight space deployment [74, 75]. Same as Arduino Uno, it is also powered with ATmega328 processor which operates at 16 MHz. the memory capacity and input/output pin count are also the same as Uno for this model [71, 74]. The only disadvantage of Arduino Nano is that it cannot connect to a number shields compatible with Arduino Uno. However, the pin headers make it easier to deploy on a breadboard. Therefore, it is not widely referred to as real-time smart monitoring applications, unless, cost-effectiveness is the prime requirement [71, 74]. Arduino Mega is the higher version of Arduino series microcontrollers with 54 input–output pins. It makes use of ATmega2560 processor clocked at 16 MHz. Other than this, it supports, 4 KB EEPROM, 8 KB RAM, and 256 KB ROM. Furthermore, Arduino Mega also support 16 analog channels along with 15 PWM channels.

Raspberry based MCUs are also available in four different versions where most of the researchers preferred using Raspberry Pi. The basic module Raspberry Pi Zero usually comes with 512 MB RAM and 1 GHz single-core ARM11 [76]. The Raspberry Pi 2B is available with 1 GB RAM and is supported by 900 MHz Quad-Core ARM Cortex-A7. The Raspberry Pi 3B is a little higher version with 1 GB RAM and a 1.2 GHz Quad-Core 64-bit ARM Cortex A53 processor. Researchers have also used Raspberry Pi3B+ in the past for designing IAQ monitoring system. Other available options for MCUs to design IoT based smart pollutant measuring systems are Sunspot Module, Texas Instruments CC3200, STM32F103C8T6 (ARM), Intel Edison Board, ARM Cortex-M0, and MSP430F5529 module.

3.4 Communication and Data Access

The development of real-time monitoring systems for pollutant concentrations in the indoor environment is not limited to sensor and gateway connections. It is crucial to identify reliable communication technology to access recorded data remotely. Literature reveals that Wi-Fi is one of the most preferred choice for communication in IAQ monitoring systems [28, 28, 30, 35]. Nevertheless, other researchers also preferred using ZigBee and Bluetooth technology [39, 55, 56, 62, 66]. The main limitation of Wi-Fi for real-time measurements is its higher power consumption requirement [28]. On the other side, Bluetooth and ZigBee are rated high for their low power requirements [77]. The Wi-Fi communication is generally established using IEEE802.11 b/n/g protocol. However, IEEE 802.15.4 is recommended for ZigBee communication [78, 79]. Several researchers also worked on the MQTT protocol for its ability to support easy implementation and low power consumption [80]. The main goal of these communication technologies is to transfer recorded data to the dedicated server from the target site.

Communication technologies are required to ensure remote storage and access of data. However, researchers also need to make a reliable selection for data storage platforms. Generally, the best recommendation for IoT based monitoring is cloud servers that allow easy access to stored data from anywhere, at any time. Moreover, it is also possible to add password protection to avoid unauthorized access to environmental data for specific commercial or residential monitoring applications. ThingSpeak is the most preferred storage solution for IAQ monitoring systems. ThingSpeak allows storage of data from four different channels where each channel can contain 8 unique parameters. This is an efficient solution for large monitoring projects. Many researchers in the past also preferred using local servers, mobile internal storage, and SD cards for data storage. Other than this, IoT datastore services can be also accessed for IAQ data.

The recorded data must be accessible to the end-users and for this, researchers also need to work on a selection of the most reliable data consulting methods. The number of smartphone users is increasing over the years, it is easier to provide real-time access to environmental data via mobile apps [81]. However, the use of web servers and portals is another popular recommendation for displaying IAQ characteristics from the target environment. Several researchers prefer using an LCD with the hardware monitoring system to display pollutant concentrations on-site. Nevertheless, most of the modern age systems are made compatible with mobile apps to provide clear insights on the go. The web-based systems can be secured with login credentials to get parameter updates.

3.5 Challenges in Real-Time Implementations

Future challenges in real-time monitoring applications with IoT also include mobility, compatibility, precision, scalability, interoperability, privacy, security, reliability, availability, and performance [82]. IoT sensor networks are required to facilitate information on a real-time basis; therefore, availability is a critical concern [83]. Regular monitoring requires a high guarantee for the availability and performance of field sensors so that data can be accessed without any interruption [48]. Redundant maintenance programs must be also designed to avoid system failures on the monitoring sites. Several factors affect the performance of an IoT system such as a huge amount of data, network traffic, and heavy reliance on the internet [83, 84]. Although internet and cloud storage facilitates easy sharing of data, it is critical to establish a stable internet connection for 24×7 h monitoring [85]. In critical applications of smart environment monitoring, the reliability of the

IoT sensor network plays a critical role [86, 87]. It is not only about being able to send information about monitored data; instead, the system must be capable enough to adapt to the changing environmental conditions. The long-term usability requires sensor nodes to possess acceptable operating ranges so that they do not require frequent maintenance due to seasonal variations in the thermal conditions. The security and privacy for field data access is another crucial factor for IoT based monitoring [82]. The MCUs generally have limited storage capacity which is not enough for handling extended real-time monitoring. Therefore, data must be stored remotely using reliable techniques [88]. It is better to use advanced encryption and authentication algorithms to secure field data from unwanted access. The interoperability of sensors is another relevant concern for IoT applications. As a variety of sensors with different specifications can be connected to a single IoT network, they should facilitate desired monitoring services irrespective of the connected hardware characteristics. It is crucial to follow standardized protocols at the application and network-level to ensure interoperability in all conditions [89]. Mobility and precision management are equally important for smart IoT environments. Moreover, sensors are desired to be compatible with the surrounding conditions and available hardware and software.

The above sections provide a detailed discussion about the availability of sensors, MCUs, and communication platforms for the development of IAQ monitoring systems. However, the real-time implementation of such systems always requires additional efforts to deal with potential challenges [90]. Smart monitoring applications for urban and rural sites are greatly affected by the cost of hardware development and deployment, power requirements, and calibration requirements [25]. Another critical concern is the system failure due to several on-site issues. The cost of the sensors is a matter of concern in the initial decision-making process, especially when the system is desired to monitor environmental conditions in rural areas and low-income countries. Although the online marketplaces are loaded with a variety of sensors with competitive price ranges, the shipping cost to deliver sensors to different locations may lead to the expensive purchase. Generally, most of the thermal comfort sensors are available within the estimated price of $5 and few others may vary up to $10. Consequently, the average cost for gas sensors and PM sensors falls below $20 [16, 25].

While deciding on sensors for IAQ monitoring, the developers also need to consider the cost of additional hardware requirements, installation, setup, maintenance, and field calibration. Long-term monitoring can pose an additional challenge in terms of repeated repair and replacement of faulty products. Depending upon how many pollutants are required to be monitored from the target site, the IAQ monitoring system may need multiple sensor nodes. Based on the number of monitoring parameters, researchers also need to be careful about MCU selection and data storage requirements. Generally, the Waspmote sensor is preferred by several researchers due to its ability to monitor multiple pollutants with easily replaceable probes. Furthermore, it also provides easy installation but the main challenge for using these sensors is repeated calibration requirement. The repeated purchase of probes may also increase the overall cost of monitoring.

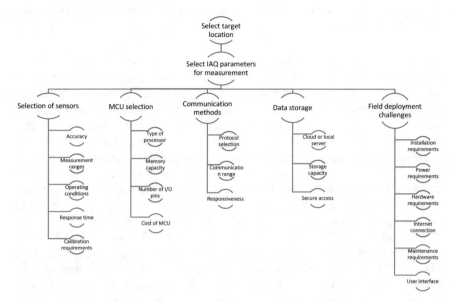

Fig. 3.2 Potential challenges required to be addressed for designing efficient IAQ monitoring systems

The sensor units must be also analyzed based on the acceptable operating conditions, measurement ranges, calibration requirements, accuracy ranges, and cost. Figure 3.2 provides quick insights into the potential challenges that are required to be addressed step by step for designing efficient IAQ monitoring systems. Moreover, real-time implementation also requires an in-depth analysis of power requirements, energy efficiency, additional hardware, and maintenance needs. Air quality plays critical role in the health and well-being of building occupants. Therefore, it is not feasible to take a risk by installing random sensing units. The IoT based IAQ monitoring systems are further required to send real-time updates to the end-users regarding pollutant concentrations. The developers need to find reliable options for generating alerts to users. SMS and email-based alerts are more useful as they can help to building occupants to take instant decisions about ventilation and control [91]. However, in rural areas and developing countries, it is also critical to establish a stable internet connection to generate such alerts on regular basis. These decisions affect the overall functionality of the IAQ monitoring system. Therefore, researchers need to make informed decisions by analyzing all associated aspects to address these challenges.

3.6   Discussions

Several studies provide evidence for the potential of IoT technology in real-time monitoring applications. IAQ monitoring is a crucial application domain for IoT technology. The manufacturers have developed several sensors for measuring different indoor air pollutants. The existing studies focused on VOCs, PM_{10}, $PM_{2.5}$, CO, CO_2, and NO_x for analyzing the quality of breathable air in the building environment. The researchers in the past have used different types of sensors for monitoring the indoor environment; they can be generally categorized into four types as dust sensors, single gas sensors, multi-gas sensors, and thermal comfort sensors. Thermal comfort parameters must be given priority for monitoring the indoor environment as they are closely associated with the overall health and comfort of the building occupants. Furthermore, the variation in temperature and humidity also affects reactions between primary and secondary pollutants. Most of the sensors in the market are available in the pre-calibrated form. However, it is crucial to conduct reliability tests before field deployment of these sensors. Moreover, before making a selection for sensors to monitor the indoor environment, researchers also need to analyze their measuring ranges, operating conditions, power requirements, and maintenance requirements.

The selection of pollutants to be measured in the target indoor environment must be made based on the activity levels. Rural homes, where the burning of solid fuels is a routine activity, can have higher concentrations of CO and CO_2 levels. Emissions of these gases are also linked to leaking chimneys, dryers, stoves, fireplaces, wood stoves, and unvented gas appliances. The presence of PM is equally high in rural and urban homes due to cooking practices, combustion, building materials, and many chemical-rich products. Several studies recommend the measurement of VOCs, CO_2, and CO as potential IAQ pollutants along with temperature and humidity in the hospital areas. However, for the school buildings, it is essential to focus on the measurement of bacteria, fungi, CO, CO_2, O_3, VOCs, and HCHO. Furthermore, the office premises can have a higher concentration of PM, CO, CO_2, VOCs, microbial compounds along with thermal comfort parameters. While making a selection of sensors for measuring these parameters, researchers also need to focus on the additional hardware required for the implementation. The sensor units are required to be connected via reliable MCUs. The existing studies in the literature recommend the use of ESP8266, Raspberry Pi, and Arduino Uno as potential solutions for measuring IAQ levels. Consequently, Wi-Fi, Zigbee, and Bluetooth are the most preferred choices for communication technologies. Most of the researchers in the past preferred using open-source platforms for designing cost-efficient IAQ monitoring systems.

For IoT based IAQ monitoring applications, researchers are also required to pay attention to deployment-related challenges such as availability of power, energy consumption, operating ranges, and maintenance cost. The system must be further able to provide instant updates to the end-users regarding critical threshold levels. The new age smart monitoring systems are required to provide an enhanced user interface via mobile apps. The web server and mobile app-based monitoring can support enhanced visualization, timely notification, and analytics of data. This strategy can help end-users to take instant actions for implementing preventive measures when pollutant thresholds cross safe levels. Nevertheless, future systems must be easy to install, cost-effective, and energy-efficient so that they can serve low-income populations. These systems can be useful to support public health and well-being.

3.7 Conclusion

This chapter presents a detailed analysis of IoT-based IAQ monitoring applications along with potential opportunities and challenges in this domain. The main objective of this chapter was to provide insights into the existing developments in this field while opening doors for the improvements. This state of the art can provide a strong foundation for future work in the field of IAQ monitoring and environmental health enhancement. On the one hand, existing studies show the great potential of IoT technology for developing smart environmental monitoring solutions. On the other hand, this in-depth analysis provides insights into the potential gaps in the field. Future researchers need to work on overcoming existing limitations while providing reliable and efficient systems for IAQ monitoring and assessment.

Most of the systems developed in the past does not follow adequate calibration procedures and reliability test before deployment of IAQ sensors on the target locations. Furthermore, several IoT based monitoring systems lack in generating notifications or triggers for alerting end-users regarding decaying pollutant concentrations in the environment. Future researchers also need to work on lowering the energy consumption of real-time monitoring systems. The accuracy of sensors and safe storage of data must be another important concern for addressing challenges in the field of IoT for the environment. This chapter provides useful insights to future researchers, policymakers, industry experts, and government authorities to take relevant actions in this domain.

References

1. J.M. Seguel, R. Merrill, D. Seguel, A.C. Campagna, Indoor air quality. Am. J. Lifestyle Med. **11**(4), 284–295 (2017). https://doi.org/10.1177/1559827616653343
2. J. Saini, M. Dutta, G. Marques, A comprehensive review on indoor air quality monitoring systems for enhanced public health. Sustain. Environ. Res. **30**(1), 6 (2020). https://doi.org/10.1186/s42834-020-0047-y
3. F. Ahmed et al., Impact of household air pollution on human health: source identification and systematic management approach. SN Appl. Sci. **1**(5), 418 (2019). https://doi.org/10.1007/s42452-019-0405-8
4. K. Apte, S. Salvi, Household air pollution and its effects on health. F1000 Res **5**, 2593 (2016). https://doi.org/10.12688/f1000research.7552.1

5. Z.A. Chafe et al., Household cooking with solid fuels contributes to ambient $PM_{2.5}$ air pollution and the burden of disease. Environ. Health Perspect. **122**(12), 1314–1320 (2014). https://doi.org/10.1289/ehp.1206340

6. T.F. Cooke, Indoor air pollutants: a literature review. Rev. Environ. Health **9**(3) (1991). https://doi.org/10.1515/REVEH.1991.9.3.137

7. L.D. Turner, S.M. Allen, R.M. Whitaker, The influence of concurrent mobile notifications on individual responses. Int. J. Hum Comput Stud. **132**, 70–80 (2019). https://doi.org/10.1016/j.ijhcs.2019.07.011

8. G.S. Graudenz, C.H. Oliveira, A. Tribess, C. Mendes, M.R.D.O. Latorre, J. Kalil, Association of air-conditioning with respiratory symptoms in office workers in tropical climate. Indoor Air **15**(1), 62–66 (2005). https://doi.org/10.1111/j.1600-0668.2004.00324.x

9. S. Sun, X. Zheng, J. Villalba-Díez, J. Ordieres-Meré, Indoor air-quality data-monitoring system: long-term monitoring benefits. Sensors **19**(19) (2019), Art. no. 19. https://doi.org/10.3390/s19194157

10. A. Kankaria, B. Nongkynrih, S.K. Gupta, Indoor air pollution in India: implications on health and its control. Indian J. Commun. Med. **39**(4), 203–207 (2014). https://doi.org/10.4103/0970-0218.143019

11. M. Krzyzanowski, J.J. Quackenboss, M.D. Lebowitz, Chronic respiratory effects of indoor formaldehyde exposure. Environ. Res. **52**(2), 117–125 (1990). https://doi.org/10.1016/S0013-9351(05)80247-6

12. U. Brunelli, V. Piazza, L. Pignato, F. Sorbello, S. Vitabile, Two-days ahead prediction of daily maximum concentrations of SO_2, O_3, PM_{10}, NO_2, CO in the urban area of Palermo, Italy. Atmos. Environ. **41**(14), 2967–2995 (2007). https://doi.org/10.1016/j.atmosenv.2006.12.013

13. C. Ghergu et al., Dealing with indoor air pollution: an ethnographic tale from urban slums in Bangalore. Int. J. Health Sci. **6**(1), 348–361 (2016)

14. Y.-S. Jun, C.-H. Jeong, K. Sabaliauskas, W. Richard Leaitch, G.J. Evans, A year-long comparison of particle formation events at paired urban and rural locations. Atmosp. Pollut. Res. **5**(3), 447–454 (2014). https://doi.org/10.5094/APR.2014.052

15. A.L. Singh, S.J. Aligarh, A comparative analysis of indoor air pollution due to domestic fuel used in rural and urban households: a case study. Transactions **35**(2), 287–298 (2013)

16. J. Saini, M. Dutta, G. Marques, Indoor air quality monitoring systems based on internet of things: a systematic review. Int. J. Environ. Res. Publ. Health **17**(14) (2020), Art. no. 14. https://doi.org/10.3390/ijerph17144942

17. K.M. Simitha, M.S. Raj, IoT and WSN based air quality monitoring and energy saving system in SmartCity project, in *2019 2nd International Conference on Intelligent Computing, Instrumentation and Control Technologies* (*ICICICT*), vol. 1 (2019), pp. 1431–1437

18. Y. Al Horr, M. Arif, M. Katafygiotou, A. Mazroei, A. Kaushik, E. Elsarrag, Impact of indoor environmental quality on occupant well-being and comfort: a review of the literature. Int. J. Sustain. Built Environ. **5**(1), 1–11 (2016). https://doi.org/10.1016/j.ijsbe.2016.03.006

19. J. Li, S.-W. Yin, G.-S. Shi, L. Wang, Optimization of indoor thermal comfort parameters with the adaptive network-based fuzzy inference system and particle swarm optimization algorithm. Math. Probl. Eng. (2017). https://www.hindawi.com/journals/mpe/2017/3075432/. Accessed 31 Dec 2020

20. F.I. Vázquez, W. Kastner, M. Kofler, Holistic smart homes for air quality and thermal comfort. Intell. Decis. Technol. **7**(1), 23–43 (2013). https://doi.org/10.3233/IDT-120149

21. M. Braik, A. Sheta, H. Al-Hiary, Hybrid neural network models for forecasting ozone and particulate matter concentrations in the Republic of China. Air Qual. Atmos. Health **13**(7), 839–851 (2020). https://doi.org/10.1007/s11869-020-00841-7

22. F.X. Ming, R.A.A. Habeeb, F.H.B. Md Nasaruddin, A.B. Gani, Real-time carbon dioxide monitoring based on IoT & cloud technologies, in *Proceedings of the 2019 8th International Conference on Software and Computer Applications*, Penang, Malaysia, Feb 2019, pp. 517–521. https://doi.org/10.1145/3316615.3316622

23. X. Yang, Q. Chen, J.S. Zhang, Y. An, J. Zeng, C.Y. Shaw, A mass transfer model for simulating VOC sorption on building materials. Atmos. Environ. **35**(7), 1291–1299 (2001). https://doi.org/10.1016/S1352-2310(00)00397-6

24. R.B. Hamanaka, G.M. Mutlu, Particulate matter air pollution: effects on the cardiovascular system. Front. Endocrinol. **9** (2018). https://doi.org/10.3389/fendo.2018.00680
25. J. Saini, M. Dutta, G. Marques, Sensors for indoor air quality monitoring and assessment through internet of things: a systematic review. Environ. Monit. Assess. **193**(2), 66 (2021). https://doi.org/10.1007/s10661-020-08781-6
26. L. Luo, Y. Zhang, B. Pearson, Z. Ling, H. Yu, X. Fu, On the security and data integrity of low-cost sensor networks for air quality monitoring. Sensors **18**(12), 4451 (2018). https://doi.org/10.3390/s18124451
27. G. Marques, J. Saini, M. Dutta, P.K. Singh, W.-C. Hong, Indoor air quality monitoring systems for enhanced living environments: a review toward sustainable smart cities. Sustainability **12**(10), 4024 (2020). https://doi.org/10.3390/su12104024
28. X. Yang, L. Yang, J. Zhang, A WiFi-enabled indoor air quality monitoring and control system: the design and control experiments, in *2017 13th IEEE International Conference on Control Automation (ICCA)*, July 2017, pp. 927–932. https://doi.org/10.1109/ICCA.2017.8003185
29. F. Lachhab, M. Bakhouya, R. Ouladsine, M. Essaaidi, Context-driven monitoring and control of buildings ventilation systems using big data and Internet of Things–based technologies. Proc. Inst. Mech. Eng. Part I J. Syst. Control Eng. **233**(3), 276–288 (2019). https://doi.org/10.1177/0959651818791406
30. N. Azmi et al., Design and development of multi-transceiver lorafi board consisting LoRa and ESP8266-Wifi communication module. IOP Conf. Ser. Mater. Sci. Eng. **318**, 012051 (2018). https://doi.org/10.1088/1757-899X/318/1/012051
31. W.-L. Hsu et al., Establishment of smart living environment control system. Sens. Mater. **32**(1), 183 (2020). https://doi.org/10.18494/SAM.2020.2581
32. A. Martín-Garín, J. A. Millán-García, A. Baïri, J. Millán-Medel, J.M. Sala-Lizarraga, Environmental monitoring system based on an open source platform and the internet of things for a building energy retrofit. Autom. Constr. **87**, 201–214 (2018). https://doi.org/10.1016/j.autcon.2017.12.017
33. N.A. Zakaria, Z. Zainal, N. Harum, L. Chen, N. Saleh, F. Azni, Wireless internet of things-based air quality device for smart pollution monitoring. Int. J. Adv. Comput. Sci. Appl. **9**(11) (2018). https://doi.org/10.14569/IJACSA.2018.091110
34. B. Vergauwen, O.M. Agudelo, R.T. Rajan, F. Pasveer, B. De Moor, Data-driven modeling techniques for indoor CO_2 estimation, in *2017 IEEE Sensors*, Oct 2017, pp. 1–3. https://doi.org/10.1109/ICSENS.2017.8234156
35. M. Rahman et al., An adaptive IoT platform on budgeted 3G data plans. J. Syst. Architect. **97**, 65–76 (2019). https://doi.org/10.1016/j.sysarc.2018.11.002
36. K.B. Kumar Sai, S. Mukherjee, H. Parveen Sultana, Low cost IoT based air quality monitoring setup using Arduino and MQ series sensors with dataset analysis. Proc. Comput. Sci. **165**, 322–327 (2019). https://doi.org/10.1016/j.procs.2020.01.043
37. E. Alexandrova, A. Ahmadinia, Real-time intelligent air quality evaluation on a resource-constrained embedded platform, in *2018 IEEE 4th International Conference on Big Data Security on Cloud (BigDataSecurity), IEEE International Conference on High Performance and Smart Computing, (HPSC) and IEEE International Conference on Intelligent Data and Security (IDS)*, May 2018, pp. 165–170. https://doi.org/10.1109/BDS/HPSC/IDS18.2018.00045
38. G. Marques, R. Pitarma, An indoor monitoring system for ambient assisted living based on internet of things architecture. Int. J. Environ. Res. Publ. Health **13**(11), 1152 (2016). https://doi.org/10.3390/ijerph13111152
39. M. Benammar, A. Abdaoui, S. Ahmad, F. Touati, A. Kadri, A modular IoT platform for real-time indoor air quality monitoring. Sensors **18**(2), 581 (2018). https://doi.org/10.3390/s18020581
40. J. Velicka, M. Pies, R. Hajovsky, Wireless measurement of carbon dioxide by use of IQRF technology. IFAC Pap Online **51**(6), 78–83 (2018). https://doi.org/10.1016/j.ifacol.2018.07.133
41. H.S. Kim et al., Development of daily PM_{10} and $PM_{2.5}$ prediction system using a deep long short-term memory neural network model, in *Aerosols/Atmospheric Modelling/Troposphere/Physics (Physical Properties and Processes), Preprint*, Mar 2019. https://doi.org/10.5194/acp-2019-268

42. J. Saini, M. Dutta, G. Marques, Particulate matter assessment in association with temperature and humidity: an experimental study on residential environment, in *Proceedings of International Conference on IoT Inclusive Life (ICIIL 2019), NITTTR Chandigarh, India*, Singapore, 2020, pp. 167–174. https://doi.org/10.1007/978-981-15-3020-3_15.

43. R.D. Brook et al., Particulate matter air pollution and cardiovascular disease. Circulation **121**(21), 2331–2378 (2010). https://doi.org/10.1161/CIR.0b013e3181dbece1

44. A.K. Gorai, P.B. Tchounwou, S. Biswal, F. Tuluri, Spatio-temporal variation of particulate matter ($PM_{2.5}$) concentrations and its health impacts in a mega city, Delhi in India. Environ. Health Insights **12** (2018). https://doi.org/10.1177/1178630218792861

45. M. Ababneh, A. AL-Manaseer, M. Hjouj Btoush, PM_{10} forecasting using soft computing techniques. Res. J. Appl. Sci. Eng. Technol. **16**, 3253–3265 (2014)

46. G. Marques, R. Pitarma, An indoor monitoring system for ambient assisted living based on internet of things architecture. Int. J. Environ. Res. Publ. Health (2017)

47. F. Pradityo, N. Surantha, Indoor air quality monitoring and controlling system based on IoT and fuzzy logic, in *2019 7th International Conference on Information and Communication Technology (ICoICT)*, July 2019, pp. 1–6. https://doi.org/10.1109/ICoICT.2019.8835246

48. R.K. Kodali, S.C. Rajanarayanan, IoT based indoor air quality monitoring system, in *2019 International Conference on Wireless Communications Signal Processing and Networking (WiSPNET)*, Mar 2019, pp. 1–5. https://doi.org/10.1109/WiSPNET45539.2019.9032855

49. G. Marques, C.R. Ferreira, R. Pitarma, A system based on the internet of things for real-time particle monitoring in buildings. Int. J. Environ. Res. Publ. Health (2018)

50. L. Zhao, W. Wu, S. Li, Design and implementation of an IoT-based indoor air quality detector with multiple communication interfaces. IEEE Internet Things J. **6**(6), 9621–9632 (2019). https://doi.org/10.1109/JIOT.2019.2930191

51. J. Saini, M. Dutta, G. Marques, Indoor air quality monitoring with IoT: predicting PM_{10} for enhanced decision support, in *2020 International Conference on Decision Aid Sciences and Application (DASA)*, Sakheer, Bahrain, Nov 2020, pp. 504–508. https://doi.org/10.1109/DASA51403.2020.9317054

52. H. Kabrein, M.Z.M. Yusof, A.M. Leman, Progresses of filtration for removing particles and gases pollutants of indoor; limitations and future direction; review article. ARPN J. Eng. Appl. Sci. **11**(6), 7 (2016)

53. M. Muladi, S. Sendari, T. Widiyaningtyas, Real time indoor air quality monitoring using internet of things at university, in *2018 2nd Borneo International Conference on Applied Mathematics and Engineering (BICAME)*, Dec 2018, pp. 169–173. https://doi.org/10.1109/BICAME45512.2018.1570509614

54. G. Marques, R. Pitarma, A cost-effective air quality supervision solution for enhanced living environments through the internet of things. Electronics, **8**(2) (2019), Art. no. 2. https://doi.org/10.3390/electronics8020170

55. Z. Tu, C. Hong, H. Feng, EMACS: design and implementation of indoor environment monitoring and control system, in *2017 IEEE/ACIS 16th International Conference on Computer and Information Science (ICIS)*, May 2017, pp. 305–309. https://doi.org/10.1109/ICIS.2017.7960010

56. Z. Idrees, Z. Zou, L. Zheng, Edge computing based IoT architecture for low cost air pollution monitoring systems: a comprehensive system analysis, design considerations & development. Sensors **18**(9), 3021 (2018). https://doi.org/10.3390/s18093021

57. M.M. Ahmed, S. Banu, B. Paul, Real-time air quality monitoring system for Bangladesh's perspective based on internet of things, in *2017 3rd International Conference on Electrical Information and Communication Technology (EICT)*, Dec. 2017, pp. 1–5. https://doi.org/10.1109/EICT.2017.8275161.

58. M. Taştan, H. Gökozan, Real-time monitoring of indoor air quality with internet of things-based E-nose. Appl. Sci. **9**(16) (2019), Art. no. 16. https://doi.org/10.3390/app9163435

59. S.C. Folea, G.D. Mois, Lessons learned from the development of wireless environmental sensors. IEEE Trans. Instrum. Meas. **69**(6), 3470–3480 (2020). https://doi.org/10.1109/TIM.2019.2938137

60. G. Chiesa, S. Cesari, M. Garcia, M. Issa, S. Li, Multisensor IoT platform for optimising IAQ levels in buildings through a smart ventilation system. Sustainability **11**(20), (2019), Art. no. 20. https://doi.org/10.3390/su11205777

61. J. Saini, M. Dutta, G. Marques, Internet of things based environment monitoring and PM_{10} prediction for smart home, in *2020 International Conference on Innovation and Intelligence for Informatics, Computing and Technologies (3ICT)*, Sakheer, Bahrain, Dec 2020, pp. 1–5. https://doi.org/10.1109/3ICT51146.2020.9311996

62. G. Marques, I.M. Pires, N. Miranda, R. Pitarma, Air quality monitoring using assistive robots for ambient assisted living and enhanced living environments through internet of things. Electronics **8**(12) (2019), Art. no. 12. https://doi.org/10.3390/electronics8121375

63. J. Choi, J.S. Park, S.J. Chang, H.R. Lee, Multi-purpose connected electronic nose system for health screening and indoor air quality monitoring, *2017 International Conference on Information Networking (ICOIN)*

64. N. Afshar-Mohajer et al., Evaluation of low-cost electro-chemical sensors for environmental monitoring of ozone, nitrogen dioxide, and carbon monoxide. J. Occup. Environ. Hyg. **15**(2), 87–98 (2018). https://doi.org/10.1080/15459624.2017.1388918

65. F. Salamone, L. Belussi, L. Danza, M. Ghellere, I. Meroni, Design and development of nEMoS, an all-in-one, low-cost, web-connected and 3D-printed device for environmental analysis. Sensors (Basel) **15**(6), 13012–13027 (2015). https://doi.org/10.3390/s150613012

66. F. Salamone, L. Belussi, L. Danza, T. Galanos, M. Ghellere, I. Meroni, Design and development of a nearable wireless system to control indoor air quality and indoor lighting quality. Sensors **17**(5) (2017), Art. no. 5. https://doi.org/10.3390/s17051021

67. G. Marques, R. Pitarma, Monitoring health factors in indoor living environments using internet of things, in *Recent Advances in Information Systems and Technologies* (Cham, 2017), pp. 785–794. https://doi.org/10.1007/978-3-319-56538-5_79

68. M. Karami, G.V. McMorrow, L. Wang, Continuous monitoring of indoor environmental quality using an Arduino-based data acquisition system. J. Build. Eng. **19**, 412–419 (2018). https://doi.org/10.1016/j.jobe.2018.05.014

69. Q.P. Ha, S. Metia, M.D. Phung, Sensing data fusion for enhanced indoor air quality monitoring. IEEE Sens. J., 1 (2020). https://doi.org/10.1109/JSEN.2020.2964396

70. G. Marques, N. Miranda, A. Kumar Bhoi, B. Garcia-Zapirain, S. Hamrioui, I. de la Torre Díez, Internet of things and enhanced living environments: measuring and mapping air quality using cyber-physical systems and mobile computing technologies. Sensors **20**(3) (2020), Art. no. 3. https://doi.org/10.3390/s20030720

71. C. Doukas, *Building internet of things with the Arduino.* S.l.: CreateSpace (2012)

72. W. Kunikowski, E. Czerwiński, P. Olejnik, J. Awrejcewicz, An overview of ATmega AVR microcontrollers used in scientific research and industrial applications. PAR **19**, 15–20 (2015). https://doi.org/10.14313/PAR_215/15

73. R.H. Sudhan, M.G. Kumar, A.U. Prakash, S.A.R. Devi, P. Sathiya, ARDUINO ATMEGA-328 microcontroller. Int. J. Innov. Res. Electr. Electron. Instrum. Control Eng. **3**(4), 27–29 (2015). https://doi.org/10.17148/IJIREEICE.2015.3406

74. K.S. Kaswan, S.P. Singh, S. Sagar, Role of Arduino in real world applications. Int. J. Sci. Technol. Res. **9**(01), 4 (2020)

75. L. Louis, Working principle of arduino and using it as a tool for study and research. IJCACS **1**(2), 21–29 (2016). https://doi.org/10.5121/ijcacs.2016.1203

76. E. Upton, G. Halfacree, *Raspberry Pi User Guide*, 4th edn. (Wiley, Chichester, 2016)

77. C. Gomez, J. Oller, J. Paradells, Overview and evaluation of bluetooth low energy: an emerging low-power wireless technology. Sensors **12**(9), 11734–11753 (2012). https://doi.org/10.3390/s120911734

78. N.V.R. Kumar, C. Bhuvana, S. Anushya, Comparison of ZigBee and bluetooth wireless technologies-survey, in *2017 International Conference on Information Communication and Embedded Systems (ICICES)*, Feb 2017, pp. 1–4. https://doi.org/10.1109/ICICES.2017.8070716

79. T. Alhmiedat, G. Samara, A low cost ZigBee sensor network architecture for indoor air quality monitoring. Int. J. Comput. Sci. Inf. Secur. (IJCSIS) **15**(1), 5 (2017)

80. N. Naik, Choice of effective messaging protocols for IoT systems: MQTT, CoAP, AMQP and HTTP, in *2017 IEEE International Systems Engineering Symposium* (*ISSE*), Vienna, Austria, Oct 2017, pp. 1–7. https://doi.org/10.1109/SysEng.2017.8088251

81. K.L. Fortuna et al., Smartphone ownership, use, and willingness to use smartphones to provide peer-delivered services: results from a national online survey. Psychiatr. Q. **89**(4), 947–956 (2018). https://doi.org/10.1007/s11126-018-9592-5

82. A.A.G.-E. Ahmed, Benefits and challenges of internet of things for telecommunication networks. Telecommun. Netw. Trends Dev. (2019). https://doi.org/10.5772/intechopen.81891

83. B.N. Silva, M. Khan, K. Han, Internet of things: a comprehensive review of enabling technologies, architecture, and challenges. IETE Tech. Rev. **35**(2), 205–220 (2018). https://doi.org/10.1080/02564602.2016.1276416

84. I. Mashal, O. Alsaryrah, T.-Y. Chung, C.-Z. Yang, W.-H. Kuo, D.P. Agrawal, Choices for interaction with things on internet and underlying issues. Ad Hoc Netw. **28**, 68–90 (2015). https://doi.org/10.1016/j.adhoc.2014.12.006

85. A.R. Biswas, R. Giaffreda, IoT and cloud convergence: opportunities and challenges, in *2014 IEEE World Forum on Internet of Things* (*WF-IoT*), Mar 2014, pp. 375–376. https://doi.org/10.1109/WF-IoT.2014.6803194

86. N. Maalel, E. Natalizio, A. Bouabdallah, P. Roux, M. Kellil, Reliability for emergency applications in internet of things, in *2013 IEEE International Conference on Distributed Computing in Sensor Systems*, May 2013, pp. 361–366. https://doi.org/10.1109/DCOSS.2013.40

87. M. Mircea, M. Stoica, B. Ghilic-Micu, Using cloud computing to address challenges raised by the internet of things, in *Connected Environments for the Internet of Things: Challenges and Solutions*, ed. by Z. Mahmood (Springer International Publishing, Cham, 2017), pp. 63–82

88. A. Al-Fuqaha, M. Guizani, M. Mohammadi, M. Aledhari, M. Ayyash, Internet of things: a survey on enabling technologies, protocols, and applications. IEEE Commun. Surv. Tutor. **17**(4), 2347–2376 (2015). https://doi.org/10.1109/COMST.2015.2444095

89. I. Ishaq et al., IETF standardization in the field of the internet of things (IoT): a survey. J. Sens. Actuator Netw. **2**(2) (2013), Art. no. 2. https://doi.org/10.3390/jsan2020235

90. S.B. Baker, W. Xiang, I. Atkinson, Internet of things for smart healthcare: technologies, challenges, and opportunities. IEEE Access **5**, 26521–26544 (2017). https://doi.org/10.1109/ACCESS.2017.2775180

91. S. Böhm, H. Driehaus, M. Wick, Contextual push notifications on mobile devices: a pre-study on the impact of usage context on user response, in *Mobile Web and Intelligent Information Systems*, vol. 11673, ed. by I. Awan, M. Younas, P. Ünal, M. Aleksy (Springer International Publishing, Cham, 2019), pp. 316–330

Chapter 4
Predicting Indoor Air Quality: Integrating IoT with Artificial Intelligence

4.1 Introduction

Air pollution can be defined as a dangerous mixture of natural and artificial substances that leads to a critical impact on human health and well-being [1]. The rise in the pollutant concentration above safe threshold levels is critical for specific groups including people with disabilities, the elderly, infants, and women that spend most of their time indoors [2]. Numerous researchers in the past have found a close association of indoor air pollution (IAP) with chronic health problems such as respiratory disease, asthma attacks, lung cancer, cardiovascular disease, and cancer [3–5]. Poor indoor air quality levels are also responsible for premature deaths, reporting 3.2 million deaths annually worldwide [6]. The trends of air pollution go worst in low-income countries where residents have the least regulation over pollutant emissions [7]. However, the cases are also rising high in the developed countries from the past few years due to increased industrial activities, poor quality building materials, and chemical-rich products [8]. In the USA, almost 60,000 premature deaths per year are linked to poor air quality levels [9]. It is the main cause behind reduced productivity levels for office employees and also causes higher absenteeism at the workplace. The business industry experiences huge losses due to health issues and productivity loss caused by poor IAQ levels [10]. Environmental Protection Agency (EPA) reports that pollutant concentrations in the indoor environment can sometimes become almost 100 times higher than in outdoor areas [11]. These conditions can be more harmful to the people who spend more time indoors, especially persons with disabilities, children, and the elderly. Therefore, IAP is ranked among the top five critical environmental health concerns for public health and well-being at a global level [11].

IAP levels are greatly influenced by the use of chemical substances, human activities, and outdoor pollutant concentrations [12]. The rise of air pollutant concentrations above safe threshold levels can impact economic, physical, and biological systems [13]. Experts advise following preventive measures to deal with the adverse effects. Scientific studies reveal that human beings can take adequate actions to deal

J. Saini et al., *Internet of Things for Indoor Air Quality Monitoring*,
SpringerBriefs in Computational Intelligence,
https://doi.org/10.1007/978-3-030-82216-3_4

with the consequences of air pollution if they get prior insights about changing pollutant concentrations. Therefore, the forecasting of IAQ pollutant concentrations has become a significant domain of work for future researchers. As compared to the monitoring of IAQ levels, the prediction is a quick and inexpensive way to analyze environmental health conditions [14]. Several researchers have developed mechanistic models for predicting IAQ [14]. These models are generally based upon the underlying mechanisms of transportation of indoor air pollutants. Literature provides evidence to several mechanistic models for forecasting concentration of particulate matter (PM) [15–18], inorganic compounds such as radon [19–22], and volatile organic compounds (VOCs) [23, 24]. The main advantage of the above-mentioned mechanistic models is their ability to work effectively at the design stage of buildings so that a clear understanding of indoor concentrations can be developed ahead of time. It can further help to build contractors to choose low-emission materials to lead effective designs. However, as mechanistic models require complex input data to work, they cannot provide reliable results when building occupants interact with the target indoor environment including window openings and indoor sources [14]. Therefore, experts recommend using statistical models as an alternative approach for forecasting pollutant concentrations in the indoor environment. The main idea behind these models is to relate measured pollutant concentrations with the questionnaire data and subsequently apply obtained information to estimate pollutant concentrations levels in the new, unfamiliar environments [25]. The measured pollutant concentrations in the specific buildings can be related to different indoor and outdoor parameters with the help of statistical models. Furthermore, pollutant concentrations can be predicted using those additional parameters. Although mechanistic models are observed to be more trustworthy for field measurements, statistical models are highly useful when the specific dynamic variations or mechanisms are not established, or when a system includes large datasets.

Recent studies show extensive use of Artificial Intelligence (AI) based models, especially machine learning in combination with statistical models to predict atmospheric pollutant concentrations in the outdoor environment [26–30]. Other than this, AI algorithms are also used to predict building energy efficiency and thermal comfort [31–33]. Commonly used models in the previous studies are regression models, decision trees, support vector machine (SVM), partial least squares, random forests, Bayesian hierarchical models, artificial neural networks, and generalized linear models [34, 35]. However, these models can deal with non-time series data preferably. To analyze and predict time series data, it is crucial to use time series models that are based on autoregression principles [14]. The common examples of such models are the autoregressive integrated moving average model and autoregressive model. The time series models are a special type of forecasting model that uses historical profiles of the parameters to forecast future values. Although time series models are already applied by existing researchers for IAQ prediction, several opportunities, challenges, and possibilities are yet to be discussed. This chapter presents a state of the art of IAQ monitoring and prediction systems. The main goal is to identify opportunities for combining the potential of IoT with AI to address the concerns associated with IAQ.

4.2 Artificial Intelligence: An Overview

AI was introduced first in the year 1956 by John McCarthy [36]. It was originally defined as a branch of science and engineering for developing intelligent machines, mainly intelligent computer programs [37]. The main goal is to use the abilities of computers to understand human intelligence. However, AI is not confined to biologically observable systems only, instead, these algorithms can be applied to a variety of applications in real-world scenarios [38]. The intelligence of these computationally efficient models involves several aspects. It is the ability to interact with the real world to understand, perceive, and act intelligently [38]. Speech recognition, image understanding, and synthesis are some domains where intelligent systems have been developed. Other than this, AI also focuses on planning, reasoning, learning, and adoption. The machines can perform intelligent operations and do introspective rational only when they are supported by extensive databases associated with the application domain. The existing information is fed in the form of rules so that machines can process details adequately and find relevant solutions for humans [39].

AI is widely recognized as an integral branch of computer science that deals with the design and study of intelligent agents for maximizing the chances of success in different applications [40]. AI machines learn from past experiences and reasoning while presenting an enhanced scope for decision making and quick response. These systems are also able to make decisions based on certain priorities while handling ambiguity and complexity more efficiently [41]. AI is concerned with the implementation of computer systems for solving problems that are otherwise difficult to handle for human beings [42]. Such problems can involve natural tasks or highly complex data. AI makes use of extensive knowledge about the application domain. Therefore, AI research and development mainly focuses on problems of knowledge representation, acquisition, and use. Furthermore, the implementation of knowledge-based systems is the most active field of research in the scientific world. Currently, AI is being used for solving problems associated with several domains including finance, bank, industry, environment, and medicine [43].

The involvement of AI-driven software solutions and smart applications is continuously evolving in our routine life. Tech experts consider AI as a modern solution for promising flexible lifestyles. However, to develop a better understanding of AI, it is first critical to know about different subdomains of AI:

- **Machine learning**: This technology is used to train machines to make decisions and inferences by utilizing past experiences and learnings. By understanding the data patterns, intelligent machines can make better decisions about handling a variety of tasks in the application domain. Therefore, machine learning algorithms are widely used in data science to make crucial decisions [44].
- **Deep learning**: This is the most advanced form of AI in which machines process information or data input through multiple layers for accurate prediction and classification of outcomes. Following the core nature of AI, the deep learning algorithms also use historical data for making in-depth analysis [45].

- **Neural networks**: As the name indicates, the functionality of neural networks is similar to the activities of neurons inside the human brain. Neural networks are used for many modern-age classification and forecasting problems [46].
- **Natural language processing**: It is the technique of making machines capable enough to read, understand and interpret human languages. Natural language processing plays an essential role in establishing solid interaction between humans and machines [47].
- **Computer vision**: Computer vision plays the most important role in image recognition techniques. This technology guides machines to learn, process, and classify a set of images [48].
- **Fuzzy logic**: Fuzzy logic is defined as a method of reasoning that is inspired by human reasoning abilities. Instead of following digital yes and no, the fuzzy systems analyze intermediate possibilities to make decisions regarding field applications [49].

Forecasting applications in the modern world are commonly developed using AI-based algorithms. Several researchers in the past have used statistical models for dealing with data collected from weather balloons, deep-space satellites, radar systems, weather monitoring systems, environmental analytics, and IoT based sensor networks. However, the increasing datasets and continuously evolving atmospheric conditions may cause fluctuation in the accuracy of forecasting models, especially when systems are expected to work for extended periods. In such applications, AI algorithms can provide enhanced reliability for forecasting accuracy with their ability to learn from past experiences.

4.3 IAQ Monitoring and Prediction Systems: State of the Art

Several studies display the potential of IoT for designing smart environmental monitoring systems [50]. These sensor networks can collect real-time data from the field for an easy understanding of changing pollutant concentration patterns. This is the main reason why experts recommend using IoT as a potential technology for developing smart villages, smart cities, and smart home projects [51]. However, smart monitoring is not enough to improve public health and well-being. As the pollutant concentrations at residential and commercial buildings may get worst at any time, forecasting environmental conditions is a critical concern for improving the building environment. AI can handle the forecasting needs to manage air quality conditions in the building environment. The prior updates about changing pollutant concentrations and possible rise above safe threshold levels can help the building occupants to take relevant preventive actions ahead of time. The framework of IAQ monitoring and prediction systems designed with integration of IoT and AI is given in Fig. 4.1.

The research communities in the past have analyzed the potential of AI for developing smart IAQ monitoring and prediction systems. The main goal is to collect field

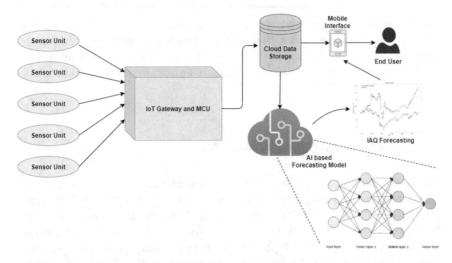

Fig. 4.1 Framework of IAQ monitoring system based on IoT and AI

data from real-time IoT based monitoring systems and process it to make predictions for future conditions. This vision has opened doors for the implementation of ambient intelligence in smart building environments that can further contribute to enhanced living. The state of the art reveals that IoT based monitoring systems can measure multiple pollutants from the target environment and store them at local or cloud-based storage systems. The data from these pre-defined storage platforms can be further utilized for training AI-based algorithms. The forecasting models analyze and understand the pattern of field data which is fed as input and predict future conditions for the target variable. The existing researchers proposed design of IAQ prediction systems based on several AI algorithms such as artificial neural network (ANN) [26, 30], recurrent neural network (RNN) [52, 53], gated recurrent neural networks (GRNN) [28], long-short term memory networks (LSTM) [54, 55], multilayer perceptron (MLP) [56], linear regression [30], Bayesian network [57, 58], adaptive neuro-fuzzy inference systems (ANFIS) [28, 59], genetic algorithm (GA) [60, 61] and SVM [62].

Artificial intelligence-based models, mainly ANN or other non-linear approaches provide an efficient, flexible, and less assumption dependent modeling solution for environmental systems. Researchers have already used different forms of neural network for analyzing and forecasting different air pollutants such as PM_{10} [63], $PM_{2.5}$ [64, p. 5], O_3 [65], NO_X [66], SO_X [66] and CO_X [67]. On the other hand, researchers recommend using SVM for environmental applications due to its structural risk minimization principle. This algorithm behaves better for minimizing errors in the training data [14]. The combination of ANN with SVM and GA provides enhanced abilities to forecast pollutant concentrations in the indoor environment. The performance of forecasting models can be further enhanced using neuro-fuzzy systems due to their ability to deal with uncertainty in input data. However, the selection of the most adequate technique for designing a prediction model is mainly dependent on different factors such as the type of data, the dependency of input variables, and the amount of data. Furthermore, the selection of the most reliable and relevant input parameters for training AI-based forecasting models is another crucial concern for researchers.

4.3.1   Selection of Input Parameters

The existing researchers used IoT based air pollutant data and meteorological information for modeling IAQ levels. However, as several variables affect the pollutant concentrations in the building environment, it is important to be careful while deciding the input space consisting of few dominant variables. Different techniques are proposed for handling this problem such as statistical modeling [68], cross-correlation analysis (CCA) [69], principle component analysis (PCA) [70], Random Forest (RF), and rough set theory (RST). The main goal is to reduce the number of input variables in such a manner that system complexity can be reduced without compromising for valuable data. The CCA is considered the simplest approach to understand the relationship between parameters involved in multiple time series data [69]. This linear data analysis technique can assist in the selection number of input variables while providing better generalization performance for the prediction target variable [71]. Another commonly used approach is PCA that is capable enough to extract relevant information out of highly complex datasets. This non-parametric method reduces the dimensionality of the complex input dataset while identifying the most meaningful variables that are represented as principal components [72]. However, this approach is suitable only for two-dimensional data where variables have a linear relationship with each other [14].

Several researchers have also used improved or modified versions of PCA to handle multi-dimensional or non-linear data; some of these are hierarchical PCA [73], nonlinear PCA [74], dynamic PCA [75], and online adaptive PCA [76]. Literature provides evidence for use of PCA for predicting environmental variables such as $PM_{2.5}$, PM_{10} [77], O_3 [30], and air quality index (AQI) [78]. Furthermore, RF helps to compute the contribution of every variable in the high dimensional dataset to lead the decision-making process. This technique is already used by researchers for analyzing air pollutant concentrations including O_3 [79], CO [79], SO_2 [79, 80], NO_2 [79], PM_{10} [79], and $PM_{2.5}$ [81]. Another option in the literature for analyzing input variables is RST that can determine even the smallest attribute of the variable and its contribution to the original dataset. This technique is mainly used in pattern recognition, knowledge discovery, and medical diagnostic databases [82–84]. Other than this, researchers also need to deal with the missing data problem which is common in data obtained from IoT based real-time monitoring systems [85]. The accuracy and precision of the forecasting model are highly affected by techniques used for missing value analysis and feature selection.

4.3.2 Existing AI Methods for IAQ Prediction

ANN, LSTM, fuzzy logic, and neuro-fuzzy systems are widely used for environmental monitoring. Several researchers have also created hybrid approaches by combining the available methods to improve the performance of forecasting models. Hrust et al. [67] employed ANN for forecasting four different air pollutants (PM_{10}, CO, O_3, and NO_2) whereas Pawlak et al. [65] proposed an ANN model for predicting the hourly mean of O_3 concentrations in the urban and rural areas of central Poland. Biancofiore et al. [63] used RNN for predicted daily average concentrations of PM_{10} and $PM_{2.5}$ levels using metrological data. Corani [69] have compared the performance of pruned neural network against the FFNN model for predicting PM_{10} and O_3 concentrations in the target environment. MLP-NN model in combination with Bayesian learning scheme was used by Anderetta et al. [86]. This model helped to forecast SO_2 concentrations in an industrial area while highlighting that SO_2 levels rise above safe threshold levels only during rare episodes. Gualtieri et al. [87] used hourly concentrations of PM_{10}, $PM_{2.5}$, SO_2, NO_X, and essential metrological parameters for predicting short-term hourly concentrations of PM_{10} with the help of the ANN model. Several researchers in the past used MLP-NN methods for forecasting IAQ parameters [88–92]; however, the focus parameters were SO_2, O_3, and PM_{10} in most cases. Although ANN models are used in numerous studies in the past, these models can show unstable performance due to their higher dependence on data. Furthermore, the network parameters such as number of epochs, type of transfer function, number of hidden layers, number of neurons, biases, and initial weights along with the amount of training data influence the overall prediction performance. These models also experience more difficulty in function approximation when input parameters or features do not possess linearly separable characteristics.

The SVM technique was initially introduced by Boser et al. for handling classification problems. However, over the years, researchers extended this technique to design regression models. Like other engineering fields, SVM is also used for environmental modeling problems. Ortiz-Garcia et al. [93] used SVM in combination with MLP-NN to predict the hourly concentration of O_3 from five different locations in Madrid, Spain. Luna et al. [94] combined SVM with ANN to predict O_3 concentrations while using O_3, CO, and NO_X as primary IAQ parameters and moisture content, temperature, solar irradiation, and wind speed as meteorological parameters. Wang et al. [95] used CO, NO_2, NO, NO_X, SO_2, and respirable suspended particle (RSP) concentrations along with five important meteorological parameters (solar radiation, outdoor temperature, indoor temperature, wind direction, and wind speed) as inputs for predicting RSP concentrations. They described the use of improved SVM with adaptive RBF-NN for designing the proposed forecasting model. Observations reveal that SVM has improved generalization capability in comparison with classical NN algorithms. However, the design of SVM models is highly dependent on the adequate selection of kernel functions. Moreover, these algorithms have a slow performance at the training phase. Several researchers have also used evolutionary neural network models with SVM to achieve enhanced prediction performance [96, 97]. Among many options, the most commonly used evolutionary computational techniques are backtracking search algorithm (BSA) [98], grey wolf optimizer (GWO) [99, 100], sine cosine algorithm (SCA) [101], cuckoo search algorithm (CSA) [102, 103], gravitational search algorithm (GSA) [104, 105], particle swarm optimization (PSO) [106, 107], differential evolution (DE) [108], bat algorithm (BA) [109] and GA [61, 110]. The evolutionary algorithms are known to tackle practical problems more efficiently while helping in the better generalization of parameters; however, the process often gets computationally expensive.

The fuzzy logic models are recommended by several researchers due to their ability to handle uncertainties and imprecisions in real-world data. The set of manually developed if–then rules make it easier to control relationships between input and output parameters. This approach is flexible, simple, and customizable with an enhanced ability to handle incomplete and imprecise data. However, it is complicated to develop rules and membership functions manually, especially when dealing with massive data [111]. Furthermore, the fuzzy systems also make analysis difficult as the received output can be interpreted in several ways. Therefore, many researchers prefer combining fuzzy logic models with neural networks to design neuro-fuzzy systems. These systems possess qualities of both network methods such as learning, parallelism, fault tolerance, adaptation, and generalization abilities. Several researchers have implemented neuro-fuzzy models for environmental modeling. Carnevale et al. [49] used a neuro-fuzzy approach to predict PM_{10} and O_3 concentrations. The authors used VOC and NO_X measurements for modeling O_3 whereas VOC, SO_X, primary PM_{10}, NO_X, and NH_3 were used for PM_{10} forecasting. Yildirim and Bayramoglu [111] used ANFIS for estimation of total suspended PM and SO_2 levels while considering relative humidity, solar radiation, temperature, precipitation, pressure, wind speed, along with previous day SO_2 as input parameters. Yeganeh et al. [112] also used the ANFIS model to forecast NO_2 concentrations in Queensland, Australia. The unique combinations of fuzzy logics and NN for IAQ modeling were also used by [113–116].

When IoT sensor networks are used for long-term monitoring applications or collect data from multiple stations, it is essential to use highly efficient models for handling the massive amount of data. This states the significance of deep learning models that can handle a sufficiently large amount of data without even requiring any pre-processing. The deep learning models consist of several hidden layers with extensive learning ability to make a better prediction. Zhou et al. [117] proposed LSTM based multi-step-ahead IAQ forecasting method. To enhance system performance, they used three different deep learning algorithms dropout neural, mini-batch gradient descent, and L2 regularization over IAQ and meteorological data inputs. Athira et al. [118] used three deep learning techniques (LSTM, RNN, and GRU) to predict PM_{10} concentrations at target environments in China where GRU was observed as the best performer. Literature reveals that researchers in the past have investigated several techniques for modeling air pollutants. Nevertheless, as environmental data has a complex data format, it is difficult to handle non-linear and complicated relationships between variables. Furthermore, the main challenge for researchers is to establish a connection between IoT based hardware systems and AI-based modeling techniques to develop a standalone smart monitoring and prediction unit.

4.4 Discussion

IoT can connect billions of physical devices and sensors to collect a massive amount of field data on regular basis. This technique is influencing the lifestyle of the current generation by a considerable level with the impact of automation. On the other side, AI can be applied to various areas of science to solve advanced issues and to take intelligent actions. Also, AI can facilitate the interaction of IoT devices with human beings to achieve flexible operations. Integration of IoT with AI can open doors to incredible advancements in the technology sector. However, to develop highly efficient systems, it is first essential to analyze the capabilities of both these technologies along with their limitations.

4.4.1 Benefits of Integrating AI and IoT

The ability of IoT technology to collect a massive amount of data and the ability of AI to process a massive amount of data opens new opportunities for scientific inventions. However, before analyzing the benefits of integrated AI and IoT systems, it is essential to focus on the individual benefits of these technologies [119]. IoT encourages active communication between devices that are connected over the same network. Smart IoT systems can act responsively to unusual conditions detected by the field sensor or device. IoT can improve patient care through smart wearable sensors that can monitor conditions such as temperature, heart rate, and weight from remote locations. IoT systems can monitor atmospheric variations using smart sensor units and this information can be further used to guide building occupants to make relevant decisions for ventilation. IoT is the most trustworthy time and money-saving technology for large automation applications. Furthermore, this technology promotes a better quality of life. On the other side, AI supports automation leading to increased productivity and resource utilization. It helps in smart decision making by coordinating in data delivery, providing forecasts, and analyzing existing trends. AI supports healthcare systems with remote monitoring and analysis ability. This technology has the potential to solve complex problems in lesser time with higher accuracy as compared to manual methods. Furthermore, it also minimizes errors and manages repetitive tasks. When IoT technology is integrated with AI, this combination boosts operational efficiency with the ability to record and process a massive amount of data. The pairing of IoT and AI can help risk prediction and system automation in several field applications. IoT and AI allow easy scalability of the system to improve performance in the target environment.

4.4.2 Challenges in Integrating AI and IoT

The combination of AI with IoT opens doors for several opportunities. However, researchers also need to deal with individual and combined challenges associated with these fields. Scalability is a big challenge in IoT systems as they are required to be equally supportive for small-scale and large-scale environments. Since IoT systems need to handle a variety of objects, devices, and sensors, interoperability is a relevant challenge for real-time implementation. IoT system software must have the ability to function with a minimum number of resources to reduce running costs. Data interpretation is a considerable challenge in large IoT networks. On the other side, developing a reliable, interactive, and the trustworthy human interface is a considerable challenge with AI. As AI is a new technology in the market, the initial investment to design AI-powered systems is high. Researchers need to find ways to reduce the implementation cost to grab new opportunities. High expectations with the AI systems are another challenge for real-time implementations. The combined issues of IoT and AI integration can be even more challenging. Since IoT is collecting data from an extended range of devices and AI processes a massive amount of information, data security is a considerable challenge in implementation. As these technologies deal with a diverse range of sources and processing elements, compatibility and complexity are major challenges in building eco-systems. The new researchers also have a lack of confidence in integrating IoT with AI to develop potential applications. Cloud attacks pose risk to IoT and AI integration in real-world applications.

Other than the basic challenges in the integration of AI and IoT, there are some domain-specific issues as well that demand the attention of the research community. While implementing a combination of IoT and AI in IAQ monitoring, researchers need to pay more attention to network infrastructure and software compatibility. The installation of IAQ expert systems in the rural areas also demands additional efforts for managing power consumption, system accuracy, reliable and user interface. As AI algorithms work on extensive data monitored by IoT systems, intermediate storage of the recorded data is a crucial concern. Other than this, it is relevant to work on reducing the cost of hardware, pre-processing of recorded parameters, feature dependency issues, and development of end-user interfacing. The main goal of integration of IoT with AI for environment management applications is to design a standalone system for helping building occupants to enjoy better health and well-being.

4.5 Conclusion

Over the years, AI have gained immense popularity for air quality modelling applications as they can handle non-linearity and complexity in data with ease. This chapter presented the state of the art for using AI with IoT for predicting air pollutant concentrations in the indoor environment. It is observed that among many other technologies, ANN is widely used by the existing researchers for predicting air pollutant concentrations in the different parts of the world. Other approaches include neuro-fuzzy systems, fuzzy logic, SVM and evolutionary algorithms. Deep learning models show great potential in handling a massive amount of data obtained from the target environment. However, several researchers in the past have experienced difficulty in feature engineering and parameter selection for building smart and standalone systems for IAQ monitoring and forecasting. Additionally, this chapter also reviewed the benefits and challenges of integrating IoT with AI while opening doors for several opportunities in the future. Observations reveal that integrated environmental monitoring systems must be designed with in-depth site-specific analysis and pollutant specific parameter adjustments. However, researchers also need to pay attention to the challenges associated with the real-time implementation of such systems. The technology integration must bring reliable results along with affordable and flexible developments.

References

1. S. Sultana, A comparative analysis of air pollution detection technique using image processing, machine learning and deep learning. Mach. Learn. 5 (2019)
2. F. Wang et al., Smart control of indoor thermal environment based on online learned thermal comfort model using infrared thermal imaging, in *2017 13th IEEE Conference on Automation Science and Engineering (CASE)*, Xi'an, Aug 2017, pp. 924–925. https://doi.org/10.1109/COASE.2017.8256221
3. S. Josyula et al., Household air pollution and cancers other than lung: a meta-analysis. Environ. Health **14** (2015). https://doi.org/10.1186/s12940-015-0001-3
4. S. Baldacci et al., Allergy and asthma: effects of the exposure to particulate matter and biological allergens. Respir. Med. **109**(9), 1089–1104 (2015). https://doi.org/10.1016/j.rmed.2015.05.017
5. R.B. Hamanaka, G.M. Mutlu, Particulate matter air pollution: effects on the cardiovascular system. Front. Endocrinol. **9** (2018). https://doi.org/10.3389/fendo.2018.00680
6. C.P. Wild, Complementing the genome with an 'exposome': the outstanding challenge of environmental exposure measurement in molecular epidemiology. Cancer Epidemiol. Biomark. Prev. **14**(8), 1847–1850 (2005). https://doi.org/10.1158/1055-9965.EPI-05-0456
7. O.P. Kurmi, K.B.H. Lam, J.G. Ayres, Indoor air pollution and the lung in low- and medium-income countries. Eur. Respir. J. **40**(1), 239–254 (2012). https://doi.org/10.1183/09031936.00190211
8. E. Uhde, T. Salthammer, Impact of reaction products from building materials and furnishings on indoor air quality—a review of recent advances in indoor chemistry. Atmos. Environ. **41**(15), 3111–3128 (2007). https://doi.org/10.1016/j.atmosenv.2006.05.082
9. J. Saini, M. Dutta, G. Marques, Indoor air quality prediction systems for smart environments: a systematic review. AIS **12**(5), 433–453 (2020). https://doi.org/10.3233/AIS-200574
10. W.J. Fisk, A.H. Rosenfeld, Estimates of improved productivity and health from better indoor environments. Indoor Air **7**(3), 158–172 (1997). https://doi.org/10.1111/j.1600-0668.1997.t01-1-00002.x
11. J.M. Seguel, R. Merrill, D. Seguel, A.C. Campagna, Indoor air quality. Am. J. Lifestyle Med. **11**(4), 284–295 (2017). https://doi.org/10.1177/1559827616653343
12. S. Capolongo, G. Settimo, Indoor air quality in healing environments: impacts of physical, chemical, and biological environmental factors on users, in *Indoor Air Quality in Healthcare Facilities*, ed. by S. Capolongo, G. Settimo, M. Gola (Springer International Publishing, Cham, 2017), pp. 1–11

13. A. Dimitriou, V. Christidou, Causes and consequences of air pollution and environmental injustice as critical issues for science and environmental education, in *The Impact of Air Pollution on Health, Economy, Environment and Agricultural Sources*, Sept 2011. https://doi.org/10.5772/17654

14. W. Wei, O. Ramalho, L. Malingre, S. Sivanantham, J.C. Little, C. Mandin, Machine learning and statistical models for predicting indoor air quality. Indoor Air **29**(5), 704–726 (2019). https://doi.org/10.1111/ina.12580

15. T. Hussein, M. Kulmala, Indoor aerosol modeling: basic principles and practical applications. Water Air Soil Pollut. Focus **8**(1), 23–34 (2008). https://doi.org/10.1007/s11267-007-9134-x

16. R. Goyal, M. Khare, Indoor air quality modeling for PM_{10}, $PM_{2.5}$, and $PM_{1.0}$ in naturally ventilated classrooms of an urban Indian school building. Environ. Monit. Assess **176**(1), 501–516 (2011). https://doi.org/10.1007/s10661-010-1600-7

17. T. Schneider et al., Prediction of indoor concentration of 0.5–4 μm particles of outdoor origin in an uninhabited apartment. Atmos. Environ. **38**(37), 6349–6359 (2004). https://doi.org/10.1016/j.atmosenv.2004.08.002

18. F. Chen, S.C.M. Yu, A.C.K. Lai, Modeling particle distribution and deposition in indoor environments with a new drift–flux model. Atmos. Environ. **40**(2), 357–367 (2006). https://doi.org/10.1016/j.atmosenv.2005.09.044

19. B.P. Jelle, Development of a model for radon concentration in indoor air. Sci. Total Environ. **416**, 343–350 (2012). https://doi.org/10.1016/j.scitotenv.2011.11.052

20. C. Dimitroulopoulou, M.R. Ashmore, M.A. Byrne, R.P. Kinnersley, Modelling of indoor exposure to nitrogen dioxide in the UK. Atmos. Environ. **35**(2), 269–279 (2001). https://doi.org/10.1016/S1352-2310(00)00176-X

21. T. van Hooff, B. Blocken, CFD evaluation of natural ventilation of indoor environments by the concentration decay method: CO_2 gas dispersion from a semi-enclosed stadium. Build. Environ. **61**, 1–17 (2013). https://doi.org/10.1016/j.buildenv.2012.11.021

22. A. Kumar, R.P. Chauhan, M. Joshi, B.K. Sahoo, Modeling of indoor radon concentration from radon exhalation rates of building materials and validation through measurements. J. Environ. Radioact. **127**, 50–55 (2014). https://doi.org/10.1016/j.jenvrad.2013.10.004

23. M. Mendez, N. Blond, P. Blondeau, C. Schoemaecker, D.A. Hauglustaine, Assessment of the impact of oxidation processes on indoor air pollution using the new time-resolved INCA-Indoor model. Atmos. Environ. **122**, 521–530 (2015). https://doi.org/10.1016/j.atmosenv.2015.10.025

24. J.C. Little, A.T. Hodgson, A.J. Gadgil, Modeling emissions of volatile organic compounds from new carpets. Atmos. Environ. **28**(2) (1993), Art. no. LBL-33318. Accessed: 24 Jan 2021 [Online]. Available: https://www.osti.gov/biblio/939275-modeling-emissions-volatile-organic-compounds-from-new-carpets

25. P. Xue, C.M. Mak, H.D. Cheung, The effects of daylighting and human behavior on luminous comfort in residential buildings: a questionnaire survey. Build. Environ. **81**, 51–59 (2014). https://doi.org/10.1016/j.buildenv.2014.06.011

26. A. Challoner, F. Pilla, L. Gill, Prediction of indoor air exposure from outdoor air quality using an artificial neural network model for inner city commercial buildings. IJERPH **12**(12), 15233–15253 (2015). https://doi.org/10.3390/ijerph121214975

27. M. Ababneh, A. Al-Manaseer, M. Hjouj Btoush, PM_{10} forecasting using soft computing techniques. Res. J. Appl. Sci. Eng. Technol. **16**, 3253–3265 (2014)

28. S. Ausati, J. Amanollahi, Assessing the accuracy of ANFIS, EEMD-GRNN, PCR, and MLR models in predicting $PM_{2.5}$. Atmos. Environ. **142**, 465–474 (2016). https://doi.org/10.1016/j.atmosenv.2016.08.007

29. M. Niu, Y. Wang, S. Sun, Y. Li, A novel hybrid decomposition-and-ensemble model based on CEEMD and GWO for short-term $PM_{2.5}$ concentration forecasting. Atmos. Environ. **134**, 168–180 (2016). https://doi.org/10.1016/j.atmosenv.2016.03.056

30. S.I.V. Sousa, F.G. Martins, M.C.M. Alvim-Ferraz, M.C. Pereira, Multiple linear regression and artificial neural networks based on principal components to predict ozone concentrations. Environ. Model. Softw. **22**(1), 97–103 (2007). https://doi.org/10.1016/j.envsoft.2005.12.002

31. J. Li, S.-W. Yin, G.-S. Shi, L. Wang, Optimization of indoor thermal comfort parameters with the adaptive network-based fuzzy inference system and particle swarm optimization algorithm. Math. Probl. Eng. (2017). https://www.hindawi.com/journals/mpe/2017/3075432/. Accessed 31 Dec 2020

32. A. Beltran, A.E. Cerpa, Optimal HVAC building control with occupancy prediction, in *Proceedings of the 1st ACM Conference on Embedded Systems for Energy-Efficient Buildings—BuildSys'14* (Memphis, Tennessee, 2014), pp. 168–171. https://doi.org/10.1145/267 4061.2674072

33. S.L. Patil, H.J. Tantau, V.M. Salokhe, Modelling of tropical greenhouse temperature by auto regressive and neural network models. Biosys. Eng. **99**(3), 423–431 (2008). https://doi.org/ 10.1016/j.biosystemseng.2007.11.009

34. H. Zhao, F. Magoulès, A review on the prediction of building energy consumption. Renew. Sustain. Energy Rev. **16**(6), 3586–3592 (2012)

35. C. Bellinger, M.S. Mohomed Jabbar, O. Zaïane, A. Osornio-Vargas, A systematic review of data mining and machine learning for air pollution epidemiology. BMC Publ. Health **17**(1), 907 (2017). https://doi.org/10.1186/s12889-017-4914-3

36. V. Rajaraman, John McCarthy—father of artificial intelligence. Resonance **19**(3), 198–207 (2014). https://doi.org/10.1007/s12045-014-0027-9

37. G. Singh, A. Vedrtnam, D. Sagar, An overview of artificial intelligence, 20 Feb 2013. https:// doi.org/10.13140/RG.2.2.20660.19840.

38. A.C. Chang, Basic concepts of artificial intelligence, in *Intelligence-Based Medicine*, ed. by A.C. Chang (Academic Press, 2020), pp. 7–22 (Chapter 1)

39. M. Oprea, L. Iliadis, An artificial intelligence-based environment quality analysis system, in *Engineering Applications of Neural Networks* (Berlin, Heidelberg, 2011), pp. 499–508. https://doi.org/10.1007/978-3-642-23957-1_55

40. Z. Allam, Z.A. Dhunny, On big data, artificial intelligence and smart cities. Cities **89**, 80–91 (2019). https://doi.org/10.1016/j.cities.2019.01.032

41. F. Jiang et al., Artificial intelligence in healthcare: past, present and future. Stroke Vasc. Neurol. **2**(4) (2017). https://doi.org/10.1136/svn-2017-000101

42. M. Labani, P. Moradi, F. Ahmadizar, M. Jalili, A novel multivariate filter method for feature selection in text classification problems. Eng. Appl. Artif. Intell. **70**, 25–37 (2018). https:// doi.org/10.1016/j.engappai.2017.12.014

43. R. Vaishya, M. Javaid, I.H. Khan, A. Haleem, Artificial Intelligence (AI) applications for COVID-19 pandemic. Diab. Metab. Syndr. **14**(4), 337–339 (2020). https://doi.org/10.1016/j. dsx.2020.04.012

44. A.K. Triantafyllidis, A. Tsanas, Applications of machine learning in real-life digital health interventions: review of the literature. J. Med. Internet. Res. **21**(4), e12286 (2019). https:// doi.org/10.2196/12286

45. G. Litjens et al., A survey on deep learning in medical image analysis. Med. Image Anal. **42**, 60–88 (2017). https://doi.org/10.1016/j.media.2017.07.005

46. N. Shahid, T. Rappon, W. Berta, Applications of artificial neural networks in health care organizational decision-making: a scoping review. PLoS ONE **14**(2), e0212356 (2019). https:// doi.org/10.1371/journal.pone.0212356

47. G. Gonzalez-Hernandez, A. Sarker, K. O'Connor, G. Savova, Capturing the patient's perspective: a review of advances in natural language processing of health-related text. Yearb. Med. Inform. **26**(01), 214–227 (2017). https://doi.org/10.15265/IY-2017-029

48. A. Voulodimos, N. Doulamis, A. Doulamis, E. Protopapadakis, Deep learning for computer vision: a brief review. Comput. Intell. Neurosci. **2018** (2018)

49. C. Carnevale, G. Finzi, E. Pisoni, M. Volta, Neuro-fuzzy and neural network systems for air quality control. Atmos. Environ. **43**(31), 4811–4821 (2009). https://doi.org/10.1016/j.atm osenv.2008.07.064

50. R. Camporese, G. Borga, N. Iandelli, A. Ragnoli, New technologies and statistics: partners for environmental monitoring and city sensing, in *Statistical Methods and Applications from a Historical Perspective*, ed. by F. Crescenzi, S. Mignani (Springer International Publishing, Cham, 2014), pp. 347–358

51. K.M. Simitha, M.S. Raj, IoT and WSN based air quality monitoring and energy saving system in SmartCity project, in *2019 2nd International Conference on Intelligent Computing, Instrumentation and Control Technologies (ICICICT)*, vol. 1 (2019), pp. 1431–1437

52. M.H. Kim, Y.S. Kim, S. Sung, C. Yoo, Data-driven prediction model of indoor air quality by the preprocessed recurrent neural networks, in *2009 ICCAS-SICE*, Aug 2009, pp. 1688–1692

53. J. Loy-Benitez, P. Vilela, Q. Li, C. Yoo, Sequential prediction of quantitative health risk assessment for the fine particulate matter in an underground facility using deep recurrent neural networks. Ecotoxicol. Environ. Saf. **169**, 316–324 (2019). https://doi.org/10.1016/j.ecoenv.2018.11.024

54. Y.-S. Chang, H.-T. Chiao, S. Abimannan, Y.-P. Huang, Y.-T. Tsai, K.-M. Lin, An LSTM-based aggregated model for air pollution forecasting. Atmos. Pollut. Res. **11**(8), 1451–1463 (2020). https://doi.org/10.1016/j.apr.2020.05.015

55. T. Xayasouk, H. Lee, G. Lee, Air pollution prediction using long short-term memory (LSTM) and deep autoencoder (DAE) models. Sustainability **12**(6), 2570 (2020). https://doi.org/10.3390/su12062570

56. S. Gündoğdu, Comparison of static MLP and dynamic NARX neural networks for forecasting of atmospheric PM_{10} and SO_2 concentrations in an industrial site of Turkey. Environ. Forensics **21**(3–4), 363–374 (2020). https://doi.org/10.1080/15275922.2020.1771637

57. V. Tomaso, L. Mario, T. Elisabetta, R. Andrea, F. Bruno, A Bayesian belief network for local air quality forecasting. Chem. Eng. Trans. **74**, 271–276 (2019). https://doi.org/10.3303/CET1974046

58. X. Xie, J. Zuo, B. Xie, T.A. Dooling, S. Mohanarajah, Bayesian network reasoning and machine learning with multiple data features: air pollution risk monitoring and early warning. Nat. Hazards (2021). https://doi.org/10.1007/s11069-021-04504-3

59. R. Wongsathan, Improvement of PM-10 forecast using ANFIS model with an integrated hotspots. Sci. Technol. Asia, 61–70 (2018)

60. D.Z. Antanasijević, V.V. Pocajt, D.S. Povrenović, M.Đ Ristić, A.A. Perić-Grujić, PM_{10} emission forecasting using artificial neural networks and genetic algorithm input variable optimization. Sci. Total Environ. **443**, 511–519 (2013). https://doi.org/10.1016/j.scitotenv.2012.10.110

61. E. Kalapanidas, N. Avouris, Feature selection for air quality forecasting: a genetic algorithm approach. AI Commun. **16**(4), 235–251 (2003)

62. Z. Ghaemi, A. Alimohammadi, M. Farnaghi, LaSVM-based big data learning system for dynamic prediction of air pollution in Tehran. Environ. Monit. Assess. **190**(5), 300 (2018). https://doi.org/10.1007/s10661-018-6659-6

63. F. Biancofiore et al., Recursive neural network model for analysis and forecast of PM_{10} and $PM_{2.5}$. Atmos. Pollut. Res. **8**(4), 652–659 (2017). https://doi.org/10.1016/j.apr.2016.12.014

64. X. Feng, Q. Li, Y. Zhu, J. Hou, L. Jin, J. Wang, Artificial neural networks forecasting of $PM_{2.5}$ pollution using air mass trajectory based geographic model and wavelet transformation. Atmos. Environ. **107**, 118–128 (2015). https://doi.org/10.1016/j.atmosenv.2015.02.030

65. I. Pawlak, J. Jarosławski, Forecasting of surface ozone concentration by using artificial neural networks in rural and urban areas in Central Poland. Atmosphere **10**(2) (2019), Art. no. 2. https://doi.org/10.3390/atmos10020052

66. D. Radojević, D. Antanasijević, A. Perić-Grujić, M. Ristić, V. Pocajt, The significance of periodic parameters for ANN modeling of daily SO_2 and NO_x concentrations: a case study of Belgrade, Serbia. Atmos. Pollut. Res. **10**(2), 621–628 (2019). https://doi.org/10.1016/j.apr.2018.11.004

67. L. Hrust, Z.B. Klaić, J. Križan, O. Antonić, P. Hercog, Neural network forecasting of air pollutants hourly concentrations using optimised temporal averages of meteorological variables and pollutant concentrations. Atmos. Environ. **43**(35), 5588–5596 (2009). https://doi.org/10.1016/j.atmosenv.2009.07.048

68. J. Saini, M. Dutta, G. Marques, Particulate matter assessment in association with temperature and humidity: an experimental study on residential environment, in *Proceedings of International Conference on IoT Inclusive Life (ICIIL 2019)* (NITTTR Chandigarh, India, Singapore, 2020), pp. 167–174. https://doi.org/10.1007/978-981-15-3020-3_15

69. G. Corani, Air quality prediction in Milan: feed-forward neural networks, pruned neural networks and lazy learning. Ecol. Model. **185**(2), 513–529 (2005). https://doi.org/10.1016/j. ecolmodel.2005.01.008

70. J. Shlens, A tutorial on principal component analysis. arXiv:1404.1100 [cs, stat], Apr 2014. Accessed: 24 Jan 2021 [Online]. Available: http://arxiv.org/abs/1404.1100

71. E. Balaguer Ballester, G. Camps i Valls, J.L. Carrasco-Rodriguez, E. Soria Olivas, S. del Valle-Tascon, Effective 1-day ahead prediction of hourly surface ozone concentrations in eastern Spain using linear models and neural networks. Ecol. Modell. **156**(1), 27–41 (2002). https://doi.org/10.1016/S0304-3800(02)00127-8

72. B.M. Wise, N.L. Ricker, D.J. Veltkamp, *Upset and Sensor Failure Detection in Multivariate Processes*, p. 41

73. J.F. MacGregor, T. Kourti, Statistical process control of multivariate processes. Control. Eng. Pract. **3**(3), 403–414 (1995). https://doi.org/10.1016/0967-0661(95)00014-L

74. M.A. Kramer, Nonlinear principal component analysis using autoassociative neural networks. AIChE J. **37**(2), 233–243 (1991). https://doi.org/10.1002/aic.690370209

75. J.V. Kresta, J.F. Macgregor, T.E. Marlin, Multivariate statistical monitoring of process operating performance. Can. J. Chem. Eng. **69**(1), 35–47 (1991). https://doi.org/10.1002/cjce.545 0690105

76. S. Wold, Exponentially weighted moving principal components analysis and projections to latent structures. Chemom. Intell. Lab. Syst. **23**(1), 149–161 (1994). https://doi.org/10.1016/ 0169-7439(93)E0075-F

77. D. Voukantsis, K. Karatzas, J. Kukkonen, T. Räsänen, A. Karppinen, M. Kolehmainen, Intercomparison of air quality data using principal component analysis, and forecasting of PM_{10} and $PM_{2.5}$ concentrations using artificial neural networks, in Thessaloniki and Helsinki. Sci. Total Environ. **409**(7), 1266–1276 (2011). https://doi.org/10.1016/j.scitotenv.2010.12.039

78. B. Mu, S. Li, S. Yuan, An improved effective approach for urban air quality forecast, in *2017 13th International Conference on Natural Computation, Fuzzy Systems and Knowledge Discovery* (*ICNC-FSKD*), July 2017, pp. 935–942. https://doi.org/10.1109/FSKD.2017.839 3403

79. R. Feng et al., Recurrent Neural Network and random forest for analysis and accurate forecast of atmospheric pollutants: a case study in Hangzhou, China. J. Clean. Prod. **231**, 1005–1015 (2019). https://doi.org/10.1016/j.jclepro.2019.05.319

80. J. Li, X. Shao, R. Sun, A DBN-based deep neural network model with multitask learning for online air quality prediction. J. Control Sci. Eng., 01 July 2019. https://www.hindawi.com/ journals/jcse/2019/5304535/. Accessed 24 Jan 2021

81. Z. Shang, T. Deng, J. He, X. Duan, A novel model for hourly $PM_{2.5}$ concentration prediction based on CART and EELM. Sci. Total Environ. **651**, 3043–3052 (2019). https://doi.org/10. 1016/j.scitotenv.2018.10.193

82. X. Peng et al., Rough set theory applied to pattern recognition of partial discharge in noise affected cable data. IEEE Trans. Dielectr. Electr. Insul. **24**, 147–156 (2017). https://doi.org/ 10.1109/TDEI.2016.006060

83. M. Angamuthu, C.M., A. Asesh, Rough set approach for an efficient medical diagnosis system. Int. J. Pharm. Technol. **7**, 8049–8060 (2015)

84. Y. Wang, L. Ma, *Feature Selection for Medical Dataset Using Rough Set Theory*, Jan 2009, pp. 68–72

85. J. Saini, M. Dutta, G. Marques, Indoor air quality monitoring with IoT: predicting PM_{10} for enhanced decision support, in *2020 International Conference on Decision Aid Sciences and Application* (*DASA*). Sakheer, Bahrain, Nov 2020, pp. 504–508. https://doi.org/10.1109/DAS A51403.2020.9317054

86. M. Andretta et al., Neural networks for sulphur dioxide ground level concentrations forecasting. NCA **9**(2), 93–100 (2000). https://doi.org/10.1007/s005210070020

87. G. Gualtieri, F. Carotenuto, S. Finardi, M. Tartaglia, P. Toscano, B. Gioli, Forecasting PM_{10} hourly concentrations in northern Italy: Insights on models performance and PM_{10} drivers through self-organizing maps. Atmos. Pollut. Res. **9**(6), 1204–1213 (2018). https://doi.org/ 10.1016/j.apr.2018.05.006

88. M. Boznar, M. Lesjak, P. Mlakar, A neural network-based method for short-term predictions of ambient SO_2 concentrations in highly polluted industrial areas of complex terrain. Atmosp. Environ. Part B Urban Atmosp. **27**(2), 221–230 (1993). https://doi.org/10.1016/0957-127 2(93)90007-S

89. J. Gómez-Sanchis, J.D. Martín-Guerrero, E. Soria-Olivas, J. Vila-Francés, J.L. Carrasco, S. del Valle-Tascón, Neural networks for analysing the relevance of input variables in the prediction of tropospheric ozone concentration. Atmosp. Environ. **40**(32), 6173–6180 (2006). https://doi.org/10.1016/j.atmosenv.2006.04.067

90. G. Spellman, An application of artificial neural networks to the prediction of surface ozone concentrations in the United Kingdom. Appl. Geogr. **19**(2), 123–136 (1999). https://doi.org/10.1016/S0143-6228(98)00039-3

91. G. de Gennaro et al., Neural network model for the prediction of PM_{10} daily concentrations in two sites in the Western Mediterranean. Sci. Total Environ. **463–464**, 875–883 (2013). https://doi.org/10.1016/j.scitotenv.2013.06.093

92. P. Perez, J. Reyes, An integrated neural network model for PM_{10} forecasting. Atmos. Environ. **40**(16), 2845–2851 (2006). https://doi.org/10.1016/j.atmosenv.2006.01.010

93. E.G. Ortiz-García, S. Salcedo-Sanz, Á.M. Pérez-Bellido, J.A. Portilla-Figueras, L. Prieto, Prediction of hourly O_3 concentrations using support vector regression algorithms. Atmosp. Environ. **44**(35), 4481–4488 (2010). https://doi.org/10.1016/j.atmosenv.2010.07.024

94. A.S. Luna, M.L.L. Paredes, G.C.G. de Oliveira, S.M. Corrêa, Prediction of ozone concentration in tropospheric levels using artificial neural networks and support vector machine at Rio de Janeiro, Brazil. Atmos. Environ. **98**, 98–104 (2014). https://doi.org/10.1016/j.atmosenv.2014.08.060

95. W. Wang, Z. Xu, J.W. Lu, Three improved neural network models for air quality forecasting. Eng. Comput. **20**(2), 192–210 (2003). https://doi.org/10.1108/02644400310465317

96. G. Grivas, A. Chaloulakou, Artificial neural network models for prediction of PM_{10} hourly concentrations, in the Greater Area of Athens, Greece. Atmos. Environ. **40**(7), 1216–1229 (2006). https://doi.org/10.1016/j.atmosenv.2005.10.036

97. X. Li, A. Luo, J. Li, Y. Li, Air pollutant concentration forecast based on support vector regression and quantum-behaved particle swarm optimization. Environ. Model. Assess. **24**(2), 205–222 (2019). https://doi.org/10.1007/s10666-018-9633-3

98. M.M. Rahman, M. Shafiullah, S.M. Rahman, A.N. Khondaker, A. Amao, Md.H. Zahir, Soft computing applications in air quality modeling: past, present, and future. Sustainability **12**(10), 4045 (2020). https://doi.org/10.3390/su12104045

99. Y. Xing, J. Yue, C. Chen, Y. Xiang, Y. Chen, M. Shi, A deep belief network combined with modified grey wolf optimization algorithm for $PM_{2.5}$ concentration prediction. Appl. Sci. **9**(18) (2019), Art. no. 18. https://doi.org/10.3390/app9183765

100. A. Saxena, S. Shekhawat, Ambient air quality classification by grey wolf optimizer based support vector machine. J. Environ. Publ. Health (2017). https://www.hindawi.com/journals/jeph/2017/3131083/. Accessed 26 Jan 2021

101. R. Li, Y. Dong, Z. Zhu, C. Li, H. Yang, A dynamic evaluation framework for ambient air pollution monitoring. Appl. Math. Model. **65**, 52–71 (2019). https://doi.org/10.1016/j.apm.2018.07.052

102. S. Kueh, K. Kuok, Forecasting long term precipitation using cuckoo search optimization neural network models. Environ. Eng. Manag. J. **17** (2018). https://doi.org/10.30638/eemj.2018.127

103. W. Sun, J. Sun, Daily $PM_{2.5}$ concentration prediction based on principal component analysis and LSSVM optimized by Cuckoo search algorithm. J. Environ. Manage. **188**, 144–152 (2016). https://doi.org/10.1016/j.jenvman.2016.12.011

104. S. Zhu, L. Yang, W. Wang, X. Liu, M. Lu, X. Shen, Optimal-combined model for air quality index forecasting: 5 cities in North China. Environ. Pollut. **243**(Pt B), 842–850 (2018). https://doi.org/10.1016/j.envpol.2018.09.025

105. S. Zhu et al., $PM_{2.5}$ forecasting using SVR with PSOGSA algorithm based on CEEMD, GRNN and GCA considering meteorological factors. Atmos. Environ. **183**, 20–32 (2018). https://doi.org/10.1016/j.atmosenv.2018.04.004

106. F.S. de Albuquerque Filho, F. Madeiro, S.M.M. Fernandes, P.S.G. de Mattos Neto, T.A.E. Ferreira, Time-series forecasting of pollutant concentration levels using particle swarm optimization and artificial neural networks. Química Nova **36**(6), 783–789 (2013). https://doi.org/10.1590/S0100-40422013000600007

107. Y. Huang, Y. Xiang, R. Zhao, Z. Cheng, Air quality prediction using improved PSO-BP neural network. IEEE Access **8**, 99346–99353 (2020). https://doi.org/10.1109/ACCESS.2020.2998145

108. Rubal, D. Kumar, Evolving differential evolution method with random forest for prediction of air pollution. Proc. Comput. Sci. **132**, 824–833 (2018). https://doi.org/10.1016/j.procs.2018.05.094

109. Q. Wu, H. Lin, A novel optimal-hybrid model for daily air quality index prediction considering air pollutant factors. Sci. Total Environ. **683**, 808–821 (2019). https://doi.org/10.1016/j.scitotenv.2019.05.288

110. H. Zhao, J. Zhang, K. Wang, Z. Bai, A. Liu, A GA-ANN model for air quality predicting, in *2010 International Computer Symposium* (*ICS2010*), Dec 2010, pp. 693–699. https://doi.org/10.1109/COMPSYM.2010.5685425

111. Y. Yildirim, M. Bayramoglu, Adaptive neuro-fuzzy based modelling for prediction of air pollution daily levels in city of Zonguldak. Chemosphere **63**(9), 1575–1582 (2006). https://doi.org/10.1016/j.chemosphere.2005.08.070

112. B. Yeganeh, M.G. Hewson, S. Clifford, A. Tavassoli, L.D. Knibbs, L. Morawska, Estimating the spatiotemporal variation of NO_2 concentration using an adaptive neuro-fuzzy inference system. Environ. Model. Softw. **100**, 222–235 (2018). https://doi.org/10.1016/j.envsoft.2017.11.031

113. S. Jain, M. Khare, Adaptive neuro-fuzzy modeling for prediction of ambient CO concentration at urban intersections and roadways. Air Qual. Atmos. Health **3**(4), 203–212 (2010). https://doi.org/10.1007/s11869-010-0073-8

114. D. Mishra, P. Goyal, A. Upadhyay, Artificial intelligence based approach to forecast $PM_{2.5}$ during haze episodes: a case study of Delhi, India. Atmos. Environ. **102**, 239–248 (2015). https://doi.org/10.1016/j.atmosenv.2014.11.050

115. J.-S. Heo, D.-S. Kim, A new method of ozone forecasting using fuzzy expert and neural network systems. Sci. Total Environ. **325**(1), 221–237 (2004). https://doi.org/10.1016/j.scitotenv.2003.11.009

116. C.-J. Chung, Y.-Y. Hsieh, H.-C. Lin, Fuzzy inference system for modeling the environmental risk map of air pollutants in Taiwan. J. Environ. Manage. **246**, 808–820 (2019). https://doi.org/10.1016/j.jenvman.2019.06.038

117. Y. Zhou, F.-J. Chang, L.-C. Chang, I.-F. Kao, Y.-S. Wang, Explore a deep learning multi-output neural network for regional multi-step-ahead air quality forecasts. J. Clean. Prod. **209**, 134–145 (2019). https://doi.org/10.1016/j.jclepro.2018.10.243

118. V. Athira, P. Geetha, R. Vinayakumar, K.P. Soman, DeepAirNet: applying recurrent networks for air quality prediction. Proc. Comput. Sci. **132**, 1394–1403. https://doi.org/10.1016/j.procs.2018.05.068

119. M.M.P. Navale, Artificial intelligence and internet of things (AIoT): opportunities and challenges. Int. J. Future Gen. Commun. Netw. **13**(3), 7 (2020)

Chapter 5
Future Directions on IoT and Indoor Air Quality Management

5.1 Introduction

Indoor environments were introduced centuries ago to provide safety from domestic animals and weather conditions. The design of the buildings was further influenced by cultural beliefs, social backgrounds, economic status, and individual preferences. However, the ongoing global industrialization and increased impact of human activities on the environment has affected living conditions in the indoor and outdoor areas. Although these developments have created a flexible lifestyle on a global scale; it has also become a matter of concern for the planet due to global warming. The adverse changes in the atmospheric conditions further introduce harmful pollutants to the breathable air that can pose a serious impact on public health and well-being.

Scientific research related to the indoor environment began in the middle of the nineteenth century when researchers analyzed the presence of harmful pollutants in the indoor spaces in Europe [1]. Max von Pettenkofer, a German professor in hygiene, was experienced in preventive medicine and modern hygiene. He made efforts to understand the association between CO_2 levels and indoor air quality. The findings show that CO_2 concentrations below 1000 ppm are acceptable for the building occupants. This rule was widely followed in many countries to develop ventilation guidelines. However, the changes in weather conditions, human lifestyle, and environmental conditions require few additional efforts for indoor air quality (IAQ) management [2]. Since the modern population spends more time indoors, it is necessary to understand the impact of the indoor environment on their health and productivity levels. Indoor air pollution (IAP) contributes to approximately 2.6% global burden of disease with 1.6 million premature deaths annually, out of which 1 million deaths belong to the children below 5 years age group [3, 4]. IAP is responsible for 28% of the total deaths in the world with 39 million disability-adjusted life years due to repeated exposure to biomass fuel in indoor environments

J. Saini et al., *Internet of Things for Indoor Air Quality Monitoring*,
SpringerBriefs in Computational Intelligence,
https://doi.org/10.1007/978-3-030-82216-3_5

[5]. Considering the alarming rate of health consequences due to IAP, World Health Organization (WHO) has announced it one of the four major environmental problems in the world [6].

Previous studies on environmental health issues in India are focused around IAP due to the burning of biomass fuels in rural, urban and semi-urban homes [4, 7–9]. However, the impact of other potential sources on IAP in the urban areas is yet to be studied. Some of the key factors contributing to poor air quality in urban houses include building materials such as wood preservatives, cement, asbestos, and volatile organic compounds generated from resins, paints, polishing materials, perfumes, and cleaning agents [10]. Other than this, IAQ levels in the office premises are also affected by the use of heaters, printers, photocopiers, improper ventilation arrangements. Furthermore, the decaying concentrations of air pollutants in the outdoor area also affect the indoor environments. There are several pollutants that can be found in the indoor air including particulate matter (PM), biological aerosols, and gases [11]. They can affect human health by a considerable level while worsening the existing health problems. Several studies reveal that IAP leaves a critical impact mostly on women and children as they spend more time indoors in rural settings and therefore, are prone to repeated exposure to biomass fuel [12]. Furthermore, pool ventilation arrangements in modern homes can also worsen the air quality conditions in the indoor environment. Since the modern population spends most of the time indoors either at home or office, the impact of IAP can be considerably large. This chapter focus on the critical health problems associated with IAQ while highlighting the need for smart IAQ management. Furthermore, it provides future directions for environmental management with advanced technology integration.

5.2 IAQ and Health: Critical Consequences

The quality of indoor air, level of exposure, and the health effects due to such exposures may vary from region to region in the world. In the developed countries, the IAP is closely linked to the furnishing materials, low ventilation rate, central heating system, repeated use of chemical products, and high prevalence of allergies. On the other side, the main cause of IAP in developing countries, especially rural areas, is the burning of biomass fuel [13]. The existing studies defined a strong association between IAP and chronic health issues such as lung cancer, chronic obstructive pulmonary disease (COPD), and acute respiratory infection (ARI) [14]. Cooking smoke can further increase coughing which is the main cause behind the spread of tuberculosis infection. WHO states that solid fuel burning for cooking and heating needs, in low-income countries, is responsible for approximately 4% global burden of disease [15]. Excess mortality rates due to lung cancer have been reported among Chinese women due to repeated exposure to the smoky coal containing high sulfur content [16]. Literature also provides evidence of adverse pregnancy outcomes, asthma attacks, tuberculosis, and cataract due to solid fuel smoke [4, 17].

The negative effects of IAP lead to 2 million premature deaths on annual basis out of which 2% die due to lung cancer, 44% due to pneumonia, and 54% due to COPD [18]. The most affected groups are younger children and women that spend most of their routine time at home. The main morbidities associated with IAP are respiratory illness such as COPD, ARI, and poor perinatal outcomes such as still-birth, low birth weight, and cancer of lung, larynx, and nasopharynx [17, 19–21]. Formaldehyde is another commonly found pollutant in rural as well as urban homes. It acts as an acute irritant on the human body with the reduction in vital capacity causing bronchitis [22]. These symptoms can further cause lung cancer and leukemia [23]. The frequent combustion of coal produces certain kinds of harmful pollutants such as toxic elements and levels of sulfur dioxide in the household environment. Very few studies have been conducted to analyze the impact of wood smoke on cardiovascular health [24, 25]. As per a report published from Guatemala, the reduction in wood smoke exposure due to wide adaption of improved chimney stoves helped in lowering systolic blood pressure by approximately 3.7 mmHg, whereas diastolic blood pressure was reduced by 3.0 mmHg. Another study was conducted in the severe cold weather of Ladakh where ventilation is kept minimum [26]. In such homes, the higher exposure to soot results in morbidities that closely resembles pneumoconiosis. In modern homes, biomass fuels are now replaced by LPG for cooking purposes [27]. However, few significant health issues are also linked to the repeated use of these clean fuels. The acute lower respiratory tract infection is the main risk factor associated with LPG.

PM is another major concern for the public health and well-being in the building environment. PM contributes to chronic bronchitis, respiratory infections, COPD, and exacerbation of COPD [28]. NO_2 and SO_2 further cause exacerbation and wheezing of asthma [28, 29]. Other than this, NO_2 is responsible for deteriorating lung functions and causing respiratory infections [30]. SO_2 also plays a significant role in cardiovascular disease and exacerbation of COPD [31]. Furthermore, repeated exposure to CO is the main cause of perinatal death and low birth weight [21]. Biomass smoke, mainly polycyclic aromatics, and metal ions can also develop a cataract, and cancers in the larynx, nasopharynx, mouth, and lungs [12]. In developed countries, smoking causes more than 80% of chronic bronchitis cases; however, in many cases, it is the contributor behind COPD and emphysema [32, 33]. However, these diseases are also reported in homes where smoking is infrequent or negligible [33]. The clinical studies show a close connection of IAP with COPD, repeated infections, and bronchiectasis [19, 30]. The prevalence of cases is equally high in other developing countries such as Nepal and Pakistan where only a few number of women smoke [34, 35].

The rising cases of asthma in many countries are also demanding attention to the role of air pollution [35]. Literature provides several consistent pieces of evidence on the triggering of asthma due to environmental tobacco smoke and air pollution [36]. In developing countries, researchers have found mixed evidence on biomass smoke as a contributing factor to asthma in adults and children [37]. Researchers conducted a study on the health and well-being of elderly women and men in the houses that use biomass fuel in routine in comparison to the households with cleaner fuels [38]. Results show a higher impact of cooking smoke on asthma in female health as compared to men. On the other side, the comparison of wood smoke and cow dung with LPG usages was associated with nuclear, cortical, and mixed cataract cases whereas biomass fuel resulted in complete or partial blindness [39, 40]. In comparison to smokers, the higher emissions of CO due to biomass fuels also results in higher carbonylhemoglobin levels. The repeated exposure to biomass fuels during pregnancy leads to a 50% increase in the risk of stillbirth and a 49% rise in the risk of low birth weight [41, 42]. A summary of commonly found indoor air pollutants, sources, and associated health effects are presented in Table 5.1.

Studies conducted in developing countries such as India, Nepal, Bangladesh, and Pakistan provide enough evidence for the contribution of IAP in increasing mortalities and morbidities [11, 20, 34, 80]. There are several financial, cultural, and social factors that influence the use of variable sources of heating and cooking in the indoor environment. Other than this, the taste of food, type of dishes, and problems with smoke also vary the user's perceptions about different alternatives [81]. However, as IAP is closely linked to severe and chronic health problems, it demands serious attention from policymakers, public health experts, and researchers. The most important concern for poor IAQ conditions is the lack of awareness and poor education regarding the use of cooking and heating systems. Most of the people in low-income countries are not aware of the threats associated with traditional fuel usage on their health and well-being [11]. Furthermore, low-income levels and unventilated homes raise the concern about the protection of children and women that spend more time indoors [82]. The industry experts are required to improve the design and functionality of cooking stoves so that they can provide the smokeless and fuel-efficient solution. Moreover, during the construction of new residential buildings and rural homes, it is relevant to design adequate ventilation arrangements [83].

5.3 Smart IAQ Management

Until recently, there was a lack of understanding regarding the variability and impli-cations of air quality. Although many researchers and scientific communities started thinking about the outdoor air quality issues including climate change and green-house gases, IAP was overlooked for years. However, as human beings in both developed and developing countries spend most of their routine time indoors, it has been ranked among the top five environmental health risks [72]. The past few decades led to rapid urbanization and industrialization at different corners of the world. With these advancements, the demands for transportation, energy, and infras-tructure development are also increasing. A number of studies show considerable decay in environmental health due to changes in the human lifestyle which is further linked to the global burden of disease [81, 84]. The breathable air is loaded with several harmful pollutants that can enter the human body through the respiratory tract and penetrate deep inside the lungs [27]. The particles can also mix in the bloodstream which further causes the development of pulmonary and cardiovascular

Table 5.1 Potential IAQ pollutants, sources, and associated health effects

Pollutants	Sources	Potential health effects	References
PM	Biomass fuel combustion, fireplaces, candles, cigarette smoking, chemical products, and kerosene heaters	Inflammatory reaction, genotoxicity, oxidative stress, carcinogenicity, mutagenicity, heart attacks, lung cancer, asthma, cardiac arrhythmias, and premature death	[43–47]
CO	Gas space heaters, unvented kerosene heaters, leaking chimneys, solid fuels	Fatigue, nausea, confusion, dizziness, headache, lack of coordination, impaired vision, and flu-like symptoms	[48–51]
NO_2	Unvented combustion appliances, kerosene heaters, tobacco smoke, gas stoves, furnaces, fireplaces, coal, wood, and oil burning	Reduced lung function, respiratory infections, asthma, difficulty in breathing, allergic rhinitis, and COPD	[52–56]
SO_2	Inadequately vented gas appliances, tobacco smoke, kerosene heaters, coal stoves, malfunctioning chimneys, and wood burning	Irritation in skin, lungs, throat, nose, and eyes; coughing, chronic bronchitis, asthma, mucus secretion, and cardiovascular disease	[31, 57–59]
CO_2	Cloth dryers, furnaces, water heaters, grills, ovens, wood stoves, fireplaces	Headaches, restlessness, dizziness, difficulty breathing, tiredness, sweating, increased heart rate, convulsions, asphyxia, elevated blood pressure, and coma	[60–63]
VOC	Hobby supplies, stored fuels, air fresheners, moth repellents, disinfectants, cleaners, aerosol sprays, wood preservatives, paints, and building materials	Irritation in throat, nose, and eyes; headaches, shortness of breath, nausea, fatigue, dizziness, skin problems, damage to kidney, liver, and central nervous system	[64–67]
O_3	Disinfecting devices, air purifiers, outdoor sources, photocopy machines, laser printers,	Decreased respiratory functions, lung damage, asthma, DNA damage, cancer, systematic inflammation, cardiovascular disease	[68–71]
Biological pollutants	Pollens coming from plants, mites, insects and animals, cockroaches, pets, house dust and dampness	Wheezing, respiratory infections, allergies and asthma	[46, 72, 73]
Radon	Outdoor air, building materials and soil gas	Lung cancer	[74–77]

(continued)

Table 5.1 (continued)

Pollutants	Sources	Potential health effects	References
Heavy metals	Outdoor sources, incense burning, combustion, building materials and smoking	Cancer, respiratory illness, brain damage, cardiovascular death, carcinogenic effects	[78, 79]

disease [25]. In this scenario, it is essential to measure the surrounding environment to understand the quality of air in which we breathe. Therefore, air quality management is a significant matter of concern for building environments. There are a variety of contaminants that can be observed in the indoor air and they affect the productivity level of people who spend more time indoors. Studies reveal that improved air quality can boost cognitive functioning, employee productivity levels, and efficiency of the immune system as well [85–87].

Emission sources, meteorological conditions, and topography are potential factors that must be kept in mind while governing air pollution. Air quality monitoring must involve long term measurement trends with in-depth analysis on the variation of individual pollutant concentration levels. It is also critical to find a correlation between human activities and the changes in the pollutant threshold levels. The rapid developments in the technology sector show incredible potential to deal with the consequences associated with IAP. IoT provides remarkable performance for designing real-time monitoring systems to execute smart IAQ management concepts. Furthermore, the integration with AI can promote overall comfort and ventilation arrangement in the building environment.

The issue of IAQ measurement has been a critical aspect for communities due to the expensive equipment. However, the modern generation is more concerned about the quality of life and comfort in their daily life. The rising demands for flexibility in routines and changing lifestyle motivates the scientific community to develop smart living spaces. Smart IAQ management systems can help building occupants to get instant updates about air quality conditions. It can also guide them to adjust ventilation by directing outside air whenever needed. IoT sensors can deliver millions of data points on daily basis focusing on all essential IAQ parameters. The AI algorithms further can help to forecast the future conditions for pollutant concentrations. This integration of technologies helps to optimize the indoor environment on a real-time basis. Smart monitoring systems can also improve the productivity levels of building occupants while minimizing the cost of health care.

The smart IAQ management systems also provide remote insights to the IAQ levels while indicating pollutant concentrations via web and mobile apps. With smart devices, it is also possible to make scheduling changes for the buildings remotely with a single click. The need for smart IAQ management can be categorized into four sections: health, safety, performance, and comfort. The consequences related to health issues associated with IAQ are already documented in the previous section. Smart monitoring solutions can prevent those issues while improving the quality of life, especially for women, children, the elderly, and disabled people. Building safety is another important concern that can be addressed with air quality management. It is already discussed that building materials leave a huge impact on the quality of air circulating in the building premises. Furthermore, it is affected by the repeated use of chemical-rich products such as cleansers, perfumes. Proper monitoring of IAQ conditions can help in the selection of the right building materials for renovation and future developments. At the same time, it can help to improve ventilation arrangements depending upon specific conditions. Good air quality levels can also promote the productivity and performance of people who spend time more indoors. Smart monitoring systems can help to correct and prevent negative consequences before they occur. Ultimately, the need for smart IAQ management is related to the overall comfort and sleep quality of individuals. As environmental factors leave a great impact on our mental and physical health, adequate management can promote healthy lifestyles by all means. Other than this, regular monitoring of environmental conditions in rural and urban areas can help government authorities, public health experts, and policymakers with efficient decision making.

5.4 Opportunities

The need for smart IAQ management can be fulfilled with the integration of advanced technologies. The previous chapters have focused on the power of IoT and AI to develop efficient and reliable monitoring systems. However, this field still requires in-depth analysis and efforts from upcoming researchers. The fusion of IoT and AI enables the design of autonomous, prescriptive, and predictive systems. These technologies are considered as an integral element of today's augmented and autonomous intelligence. This collaboration can impact a variety of industries ranging from retail, manufacturing, telecommunication, healthcare, and transportation. Several studies have been already published specifically in the field of IAQ management using smarter devices [88–90]. Moreover, the evolution of the 5G landscape can provide a solid foundation for extracting the full performance of AI-powered IoT. This integration can support massive connectivity with ultra-low latency capability that can enable smart monitoring on remote sites. The development of smart IAQ sensing units is mainly focused on three things: smart devices/sensors, intelligent integration of systems, and an end to end data analysis. Although a variety of sensor units are available for measuring different indoor air pollutants, the cost, calibration, and accuracy of those sensors have been a matter of concern over the years. The sensor units must be further integrated with the microcontrollers and gateways to collect data from the remote fields. In this process, researchers and developers also need to deal with issues related to real-time implementation such as power requirements, weather changes, and system failure. As IoT systems collect a massive amount of field data, end-to-end analysis of collected information is another concern. However, AI algorithms present considerable efficiency to analyse and interpret extensive information from field IoT systems. Furthermore, developers can experience challenges in terms of design innovations and algorithmic implementations. Researchers in the past have experienced trouble in meeting Quality of Service requirements while developing smart monitoring systems due to delay, bandwidth, and latency issues. However, even after all these challenges, the integration of advanced technologies can bring several opportunities for smart monitoring needs.

The biggest potential of AI is its ability to get insights from a variable range of data. Machine learning makes it easier to identify specific patterns in air pollutant concentrations. When implemented accurately, these algorithms can also detect anomalies in the observed data due to field devices and heterogeneous sensors. Combining AI with an IoT sensor network can also enhance the operational efficiency of the system. This is because machine learning algorithms can predict the failure of equipment. At the same time, it can also predict operating conditions while identifying parameters that must be adjusted to maintain ideal outcomes. AI algorithms can analyse a massive amount of data more efficiently as compared to human eyes and indicate potential threshold shifts. For the IAQ monitoring applications, the integration of AI with IoT can also enhance risk management. These smart equipments can help residential and commercial building occupants to predict and understand air quality risks in a more accurate manner while enabling automated response in the premises. This rapid action ensures the safety of occupants while avoiding critical onsets of disease. This integration can be further useful for supporting smart villages, smart city, and smart building concepts.

The future initiatives for IAQ management must be taken at three different levels: government, research, and industrial level to create a road map for mitigating critical effects. The government agencies need to work together to design a robust plan as per country standards. It could further help in the formation of reliable policies for urban and rural areas. However, before that researchers need to work on identifying various knowledge gaps in the field of IAQ management while highlighting potential areas for interventions. It is possible to fulfil those knowledge gaps by implementing long-term monitoring systems, designing reliable protocols, doing exposure assessments, and utilizing pollutant concentration patterns into the development of air quality standards. Also, it is equally significant to launch public awareness programs to boost the knowledge base of common people in rural and urban areas. Government and policymakers can also guide people to replace biomass fuels with cleaner fuels to ensure an environmentally friendly solution. However, the main challenge is to provide affordable alternatives to users to meet their cooking and heating needs. At the industrial level, the manufacturers need to understand the demands and practices for IAQ management in building premises. The manufacturing experts can work toward the development of healthy building materials to minimize pollutant emissions. The green building concept is worth useful for modern residential and commercial premises. This approach can reduce the presence of harmful gases and particle pollutants in the indoor environment while reducing the critical impacts on human health and well-being. The industries also need to collect regular feedback from the market regarding new solutions and work on improvements to meet future challenges. In the meanwhile, researchers need to do an in-depth assessment of IAQ conditions including exposure risks, potential pollutants, and types of fuels. They also need to develop novel methods for measurement, monitoring, and forecasting of IAQ conditions in the indoor environment. The collective efforts at all stages can help in preventing critical consequences associated with public health and well-being.

5.5 Conclusion

As modern lifestyle required people to spend most of their routine time indoors, the comfortable building environments are favourable for occupants. At one side where rural residential environments are greatly affected by solid fuel combustion, the urban premises experience decay in air quality due to poor building materials and improper ventilation arrangements. Therefore, it is critical to investigate the potential of advanced technologies to prevent medical health issues associated to deteriorated air quality. The wide acceptance to smart home and smart monitoring technologies opens new doors for the environmental health management. However, the upcoming researchers and field experts need to make efforts to eliminate challenges in the real time implementation of these systems. Reliable development and successful integration of hardware and software can easily improve the indoor environment while ensuring higher comfort for the building occupants. Furthermore, the future researchers also need to work on the social, economic and functional barriers associated to smart monitoring solutions. When implemented accurately, the smart monitoring technology can change the trends for IAQ and public health.

References

1. W.G. Locher, Max von Pettenkofer (1818–1901) as a pioneer of modern hygiene and preventive medicine. Environ. Health Prev. Med. **12**(6), 238–245 (2007). https://doi.org/10.1007/BF0289 8030
2. P.P. Shrimandilkar, *Indoor Air Quality Monitoring for Human Health* (2013), p. 7
3. K.R. Smith, S. Mehta, The burden of disease from indoor air pollution in developing countries: comparison of estimates. Int. J. Hyg. Environ. Health **206**(4–5), 279–289 (2003). https://doi.org/10.1078/1438-4639-00224
4. K. Balakrishnan et al., Air pollution from household solid fuel combustion in India: an overview of exposure and health related information to inform health research priorities. Glob. Health Act. **4** (2011). https://doi.org/10.3402/gha.v4i0.5638
5. R. Goyal, M. Khare, Indoor air quality: current status, missing links and future road map for India. J. Civ. Environ. Eng. **02**(04) (2012). https://doi.org/10.4172/2165-784X.1000118
6. World Health Organization, *Burden of Disease from Household Air Pollution for 2012: Summary of Results* (World Health Organization, Geneva, Switzerland, 2015). Accessed: 04 Dec 2020 [Online]. Available: https://www.who.int/phe/health_topics/outdoorair/databases/FINAL_HAP_AAP_BoD_24March2014.pdf?ua=1
7. S. Awasthi, H.A. Glick, R.H. Fletcher, Effect of cooking fuels on respiratory diseases in preschool children in Lucknow, India. Am. J. Trop. Med. Hyg. **55**(1), 48–51 (1996). https://doi.org/10.4269/ajtmh.1996.55.48

8. K. Balakrishnan et al., Daily average exposures to respirable particulate matter from combustion of biomass fuels in rural households of southern India. Environ. Health Perspect. **110**(11), 1069–1075 (2002). https://doi.org/10.1289/ehp.021101069

9. N. Bruce, R. Perez-Padilla, R. Albalak, Indoor air pollution in developing countries: a major environmental and public health challenge. Bull. World Health Organ. **78**(9), 1078–1092 (2000)

10. P. Kumar, Footprints of airborne ultrafine particles on urban air quality and public health. J. Civ. Environ. Eng. **01**(01) (2012). https://doi.org/10.4172/2165-784X.1000e101

11. J. Saini, M. Dutta, G. Marques, A comprehensive review on indoor air quality monitoring systems for enhanced public health. Sustain. Environ. Res. **30**(1), 6 (2020). https://doi.org/10.1186/s42834-020-0047-y

12. H.W. de Koning, K.R. Smith, J.M. Last, Biomass fuel combustion and health. Bull. World Health Organ. **63**(1), 11–26 (1985)

13. E.T. Gall, E.M. Carter, C. Matt Earnest, B. Stephens, Indoor air pollution in developing countries: research and implementation needs for improvements in global public health. Am. J. Publ. Health **103**(4), e67–e72 (2013). https://doi.org/10.2105/AJPH.2012.300955

14. K.R. Smith, Indoor air pollution in developing countries and acute lower respiratory infections in children. Thorax **55**(6), 518–532 (2000). https://doi.org/10.1136/thorax.55.6.518

15. J. Sundell, On the history of indoor air quality and health. Indoor Air **14**(s7), 51–58 (2004). https://doi.org/10.1111/j.1600-0668.2004.00273.x

16. K.R. Smith, *Indoor Air Pollution in Developing Countries*, Global Environmental Health, University of California, Berkeley. http://www.kirkrsmith.org/publications/1994/08/23/199. Accessed 25 Jan 2021

17. M. Dherani, D. Pope, M. Mascarenhas, K.R. Smith, M. Weber, N. Bruce, Indoor air pollution from unprocessed solid fuel use and pneumonia risk in children aged under five years: a systematic review and meta-analysis. Bull. World Health Organ. **86**(5), 390–398C (2008). https://doi.org/10.2471/blt.07.044529

18. A. Kankaria, B. Nongkynrih, S.K. Gupta, Indoor air pollution in India: implications on health and its control. Indian J. Commun. Med. **39**(4), 203–207 (2014). https://doi.org/10.4103/0970-0218.143019

19. O.P. Kurmi, S. Semple, P. Simkhada, W.C.S. Smith, J.G. Ayres, COPD and chronic bronchitis risk of indoor air pollution from solid fuel: a systematic review and meta-analysis. Thorax **65**(3), 221–228 (2010). https://doi.org/10.1136/thx.2009.124644

20. A. Sapkota et al., Indoor air pollution from solid fuels and risk of hypopharyngeal/laryngeal and lung cancers: a multicentric case-control study from India. Int. J. Epidemiol. **37**(2), 321–328 (2008). https://doi.org/10.1093/ije/dym261

21. D.P. Pope et al., Risk of low birth weight and stillbirth associated with indoor air pollution from solid fuel use in developing countries. Epidemiol. Rev. **32**, 70–81 (2010). https://doi.org/10.1093/epirev/mxq005

22. A. Blair, R. Saracci, P.A. Stewart, R.B. Hayes, C. Shy, Epidemiologic evidence on the relationship between formaldehyde exposure and cancer. Scand. J. Work Environ. Health **16**(6), 381–393 (1990). https://doi.org/10.5271/sjweh.1767

23. E.D. Acheson, H.R. Barnes, M.J. Gardner, C. Osmond, B. Pannett, C.P. Taylor, Formaldehyde process workers and lung cancer. Lancet **1**(8385), 1066–1067 (1984). https://doi.org/10.1016/s0140-6736(84)91466-1

24. D.A. Collings, S.D. Sithole, K.S. Martin, Indoor woodsmoke pollution causing lower respiratory disease in children. Trop. Doct. **20**(4), 151–155 (1990). https://doi.org/10.1177/004947559002000403

25. R.B. Hamanaka, G.M. Mutlu, Particulate matter air pollution: effects on the cardiovascular system. Front. Endocrinol. **9** (2018). https://doi.org/10.3389/fendo.2018.00680

26. T. Norboo, M. Yahya, N.G. Bruce, J.A. Heady, K.P. Ball, Domestic pollution and respiratory illness in a Himalayan village. Int. J. Epidemiol. **20**(3), 749–757 (1991). https://doi.org/10.1093/ije/20.3.749

27. Y. Ramesh Bhat, N. Manjunath, D. Sanjay, Y. Dhanya, Association of indoor air pollution with acute lower respiratory tract infections in children under 5 years of age. Paediatr. Int. Child Health **32**(3), 132–135. https://doi.org/10.1179/2046905512Y.0000000027

28. N. Bruce, R. Perez-Padilla, R. Albalak, Indoor air pollution in developing countries: a major environmental and public health challenge. Bull. World Health Organ. 15 (2000)

29. J.L. Sublett, J. Seltzer, R. Burkhead, P.B. Williams, H.J. Wedner, W. Phipatanakul, Air filters and air cleaners: rostrum by the American academy of allergy, asthma & immunology indoor allergen committee. J. Allergy Clin. Immunol. 125(1), 32–38 (2010). https://doi.org/10.1016/j.jaci.2009.08.036

30. J.R. Armstrong, H. Campbell, Indoor air pollution exposure and lower respiratory infections in young Gambian children. Int. J. Epidemiol. 20(2), 424–429 (1991)

31. W.J. Seow et al., Indoor concentrations of nitrogen dioxide and sulfur dioxide from burning solid fuels for cooking and heating in Yunnan Province, China. Indoor Air 26(5), 776–783 (2016). https://doi.org/10.1111/ina.12251

32. H.R. Anderson, Respiratory abnormalities and ventilatory capacity in a Papua New Guinea Island community. Am. Rev. Respir. Dis. 114(3), 537–548 (1976). https://doi.org/10.1164/arrd.1976.114.3.537

33. H.R. Anderson, Respiratory abnormalities, smoking habits and ventilatory capacity in a highland community in Papua New Guinea: prevalence and effect on mortality. Int. J. Epidemiol. 8(2), 127–136 (1979). https://doi.org/10.1093/ije/8.2.127

34. M.R. Pandey, Prevalence of chronic bronchitis in a rural community of the Hill Region of Nepal. Thorax 39(5), 331–336 (1984)

35. K.A. Qureshi, Domestic smoke pollution and prevalence of chronic bronchitis/asthma in a rural area of Kashmir. Indian J. Chest Dis. Allied Sci. 36(2), 61–72 (1994)

36. L.J. Rosen, V. Myers, J.P. Winickoff, J. Kott, Effectiveness of interventions to reduce tobacco smoke pollution in homes: a systematic review and meta-analysis. Int. J. Environ. Res. Publ. Health 12(12), 16043–16059 (2015). https://doi.org/10.3390/ijerph121215038

37. D.G. Fullerton, N. Bruce, S.B. Gordon, Indoor air pollution from biomass fuel smoke is a major health concern in the developing world. Trans. R. Soc. Trop. Med. Hyg. 102(9), 843–851 (2008). https://doi.org/10.1016/j.trstmh.2008.05.028

38. V. Mishra, Effect of indoor air pollution from biomass combustion on prevalence of asthma in the elderly. Environ. Health Perspect. 111(1), 71–78 (2003). https://doi.org/10.1289/ehp.5559

39. V.K. Mishra, R.D. Retherford, K.R. Smith, Biomass cooking fuels and prevalence of tuberculosis in India. Int. J. Infect. Dis. 3(3), 119–129 (1999)

40. M. Mohan et al., India-US case-control study of age-related cataracts. India-US case-control study group. Arch. Ophthalmol. 107(5), 670–676 (1989)

41. J.M. Tielsch et al., Exposure to indoor biomass fuel and tobacco smoke and risk of adverse reproductive outcomes, mortality, respiratory morbidity and growth among newborn infants in south India. Int. J. Epidemiol. 38(5), 1351–1363 (2009). https://doi.org/10.1093/ije/dyp286

42. D.V. Mavalankar, C.R. Trivedi, R.H. Gray, Levels and risk factors for perinatal mortality in Ahmedabad, India. Bull. World Health Organ. 69(4), 435–442 (1991)

43. O. US EPA, *Particulate Matter (PM) Basics*. US EPA, 19 Apr 2016. https://www.epa.gov/pm-pollution/particulate-matter-pm-basics. Accessed 25 Jan 2021

44. Y.-H. Cheng, Measuring indoor particulate matter concentrations and size distributions at different time periods to identify potential sources in an office building in Taipei City. Build. Environ. 123, 446–457 (2017). https://doi.org/10.1016/j.buildenv.2017.07.025

45. Z. Li, Q. Wen, R. Zhang, Sources, health effects and control strategies of indoor fine particulate matter (PM2.5): a review. Sci. Total Environ. 586, 610–622 (2017). https://doi.org/10.1016/j.scitotenv.2017.02.029

46. S. Baldacci et al., Allergy and asthma: effects of the exposure to particulate matter and biological allergens. Respir. Med. 109(9), 1089–1104 (2015). https://doi.org/10.1016/j.rmed.2015.05.017

47. R.D. Brook et al., Particulate matter air pollution and cardiovascular disease. Circulation 121(21), 2331–2378 (2010). https://doi.org/10.1161/CIR.0b013e3181dbece1

48. O. US EPA, *Carbon Monoxide's Impact on Indoor Air Quality*. US EPA, 31 July 2014. https://www.epa.gov/indoor-air-quality-iaq/carbon-monoxides-impact-indoor-air-quality. Accessed 25 Jan 2021

49. D. Penney, V. Benignus, S. Kephalopoulos, D. Kotzias, M. Kleinman, A. Verrier, *Carbon Monoxide* (World Health Organization, 2010)
50. M. Fazlzadeh, R. Rostami, S. Hazrati, A. Rastgu, Concentrations of carbon monoxide in indoor and outdoor air of Ghalyun cafes. Atmos. Pollut. Res. **6**(4), 550–555 (2015). https://doi.org/10.5094/APR.2015.061
51. J.A. Raub, M. Mathieu-Nolf, N.B. Hampson, S.R. Thom, Carbon monoxide poisoning—a public health perspective. Toxicology **145**(1), 1–14, Apr 2000. https://doi.org/10.1016/S0300-483X(99)00217-6
52. O. US EPA, *Nitrogen Dioxide's Impact on Indoor Air Quality*. US EPA, 14 Aug 2014. https://www.epa.gov/indoor-air-quality-iaq/nitrogen-dioxides-impact-indoor-air-quality. Accessed 25 Jan 2021
53. D.J. Jarvis, G. Adamkiewicz, M.-E. Heroux, R. Rapp, F.J. Kelly, *Nitrogen Dioxide* (World Health Organization, 2010)
54. B.J. Smith, M. Nitschke, L.S. Pilotto, R.E. Ruffin, D.L. Pisaniello, K.J. Willson, Health effects of daily indoor nitrogen dioxide exposure in people with asthma. Eur. Respir. J. **16**(5), 879–885 (2000). https://doi.org/10.1183/09031936.00.16587900
55. H. Salonen, T. Salthammer, L. Morawska, Human exposure to ozone in school and office indoor environments. Environ. Int. **119**, 503–514 (2018). https://doi.org/10.1016/j.envint.2018.07.012
56. M. Shima, M. Adachi, Effect of outdoor and indoor nitrogen dioxide on respiratory symptoms in schoolchildren. Int. J. Epidemiol. **29**(5), 862–870 (2000). https://doi.org/10.1093/ije/29.5.862
57. O. US EPA, *Sulfur Dioxide Basics*. US EPA, 02 Jun 2016. https://www.epa.gov/so2-pollution/sulfur-dioxide-basics. Accessed 25 Jan 2021
58. M. Brauer, S. Henderson, T. Kirkham, K. Lee, K. Rich, K. Teschke, *Review of the Health Risks Associated with Nitrogen Dioxide and Sulfur Dioxide in Indoor Air* (2002)
59. J.D. Spengler, B.G. Ferris Jr., D.W. Dockery, F.E. Speizer, Sulfur dioxide and nitrogen dioxide levels inside and outside homes and the implications on health effects research, 01 May 2002. https://pubs.acs.org/doi/pdf/10.1021/es60158a013. Accessed 25 Jan 2021
60. U. Satish et al., Is CO_2 an indoor pollutant? direct effects of low-to-moderate CO_2 concentrations on human decision-making performance. Environ. Health Perspect. **120**(12), 1671–1677 (2012). https://doi.org/10.1289/ehp.1104789
61. B. Goldstein, D. Gounaridis, J.P. Newell, The carbon footprint of household energy use in the United States. PNAS **117**(32), 19122–19130 (2020). https://doi.org/10.1073/pnas.1922205117
62. D. Stokes, A. Linsay, J. Marinopoulos, A. Treloar, G. Wescott, Household carbon dioxide production in relation to the greenhouse effect. J. Environ. Manage. **40**(3), 197–211 (1994). https://doi.org/10.1006/jema.1994.1015
63. O. US EPA, *Household Carbon Footprint Calculator*. US EPA, 12 Jan 2016. https://www.epa.gov/ghgemissions/household-carbon-footprint-calculator. Accessed 25 Jan 2021
64. D.X. Ho, K.-H. Kim, J. Ryeul Sohn, Y. Hee Oh, and J.-W. Ahn, Emission rates of volatile organic compounds released from newly produced household furniture products using a large-scale chamber testing method. Sci. World J. 08 Sept 2011. https://www.hindawi.com/journals/tswj/2011/650624/. Accessed 25 Jan 2021
65. K. Rumchev, H. Brown, J. Spickett, Volatile organic compounds: do they present a risk to our health? Rev. Environ. Health **22**, 39–55 (2007). https://doi.org/10.1515/REVEH.2007.22.1.39
66. J.-Y. Chin et al., Levels and sources of volatile organic compounds in homes of children with Asthma. Indoor Air **24**(4), 403–415 (2014). https://doi.org/10.1111/ina.12086
67. O. US EPA, *Volatile Organic Compounds' Impact on Indoor Air Quality*. US EPA, 18 Aug 2014. https://www.epa.gov/indoor-air-quality-iaq/volatile-organic-compounds-impact-indoor-air-quality. Accessed 25 Jan 2021
68. G. Spellman, An application of artificial neural networks to the prediction of surface ozone concentrations in the United Kingdom. Appl. Geogr. **19**(2), 123–136 (1999). https://doi.org/10.1016/S0143-6228(98)00039-3
69. T. Wainman, J. Zhang, C.J. Weschler, P.J. Lioy, Ozone and limonene in indoor air: a source of submicron particle exposure. Environ. Health Perspect. **108**(12), 1139–1145 (2000)

49. D. Penney, V. Benignus, S. Kephalopoulos, D. Kotzias, M. Kleinman, A. Verrier, *Carbon Monoxide* (World Health Organization, 2010)
50. M. Fazlzadeh, R. Rostami, S. Hazrati, A. Rastgu, Concentrations of carbon monoxide in indoor and outdoor air of Ghalyun cafes. Atmos. Pollut. Res. **6**(4), 550–555 (2015). https://doi.org/10.5094/APR.2015.061
51. J.A. Raub, M. Mathieu-Nolf, N.B. Hampson, S.R. Thom, Carbon monoxide poisoning—a public health perspective. Toxicology **145**(1), 1–14, Apr 2000. https://doi.org/10.1016/S0300-483X(99)00217-6
52. O. US EPA, *Nitrogen Dioxide's Impact on Indoor Air Quality.* US EPA, 14 Aug 2014. https://www.epa.gov/indoor-air-quality-iaq/nitrogen-dioxides-impact-indoor-air-quality. Accessed 25 Jan 2021
53. D.J. Jarvis, G. Adamkiewicz, M.-E. Heroux, R. Rapp, F.J. Kelly, *Nitrogen Dioxide* (World Health Organization, 2010)
54. B.J. Smith, M. Nitschke, L.S. Pilotto, R.E. Ruffin, D.L. Pisaniello, K.J. Willson, Health effects of daily indoor nitrogen dioxide exposure in people with asthma. Eur. Respir. J. **16**(5), 879–885 (2000). https://doi.org/10.1183/09031936.00.16587900
55. H. Salonen, T. Salthammer, L. Morawska, Human exposure to ozone in school and office indoor environments. Environ. Int. **119**, 503–514 (2018). https://doi.org/10.1016/j.envint.2018.07.012
56. M. Shima, M. Adachi, Effect of outdoor and indoor nitrogen dioxide on respiratory symptoms in schoolchildren. Int. J. Epidemiol. **29**(5), 862–870 (2000). https://doi.org/10.1093/ije/29.5.862
57. O. US EPA, *Sulfur Dioxide Basics.* US EPA, 02 Jun 2016. https://www.epa.gov/so2-pollution/sulfur-dioxide-basics. Accessed 25 Jan 2021
58. M. Brauer, S. Henderson, T. Kirkham, K. Lee, K. Rich, K. Teschke, *Review of the Health Risks Associated with Nitrogen Dioxide and Sulfur Dioxide in Indoor Air* (2002)
59. J.D. Spengler, B.G. Ferris Jr., D.W. Dockery, F.E. Speizer, Sulfur dioxide and nitrogen dioxide levels inside and outside homes and the implications on health effects research, 01 May 2002. https://pubs.acs.org/doi/pdf/10.1021/es60158a013. Accessed 25 Jan 2021
60. U. Satish et al., Is CO_2 an indoor pollutant? direct effects of low-to-moderate CO_2 concentrations on human decision-making performance. Environ. Health Perspect. **120**(12), 1671–1677 (2012). https://doi.org/10.1289/ehp.1104789
61. B. Goldstein, D. Gounaridis, J.P. Newell, The carbon footprint of household energy use in the United States. PNAS **117**(32), 19122–19130 (2020). https://doi.org/10.1073/pnas.1922205117
62. D. Stokes, A. Linsay, J. Marinopoulos, A. Treloar, G. Wescott, Household carbon dioxide production in relation to the greenhouse effect. J. Environ. Manage. **40**(3), 197–211 (1994). https://doi.org/10.1006/jema.1994.1015
63. O. US EPA, *Household Carbon Footprint Calculator.* US EPA, 12 Jan 2016. https://www.epa.gov/ghgemissions/household-carbon-footprint-calculator. Accessed 25 Jan 2021
64. D.X. Ho, K.-H. Kim, J. Ryeul Sohn, Y. Hee Oh, and J.-W. Ahn, Emission rates of volatile organic compounds released from newly produced household furniture products using a large-scale chamber testing method. Sci. World J. 08 Sept 2011. https://www.hindawi.com/journals/tswj/2011/650624/. Accessed 25 Jan 2021
65. K. Rumchev, H. Brown, J. Spickett, Volatile organic compounds: do they present a risk to our health? Rev. Environ. Health **22**, 39–55 (2007). https://doi.org/10.1515/REVEH.2007.22.1.39
66. J.-Y. Chin et al., Levels and sources of volatile organic compounds in homes of children with Asthma. Indoor Air **24**(4), 403–415 (2014). https://doi.org/10.1111/ina.12086
67. O. US EPA, *Volatile Organic Compounds' Impact on Indoor Air Quality.* US EPA, 18 Aug 2014. https://www.epa.gov/indoor-air-quality-iaq/volatile-organic-compounds-impact-indoor-air-quality. Accessed 25 Jan 2021
68. G. Spellman, An application of artificial neural networks to the prediction of surface ozone concentrations in the United Kingdom. Appl. Geogr. **19**(2), 123–136 (1999). https://doi.org/10.1016/S0143-6228(98)00039-3
69. T. Wainman, J. Zhang, C.J. Weschler, P.J. Lioy, Ozone and limonene in indoor air: a source of submicron particle exposure. Environ. Health Perspect. **108**(12), 1139–1145 (2000)

28. N. Bruce, R. Perez-Padilla, R. Albalak, Indoor air pollution in developing countries: a major environmental and public health challenge. Bull. World Health Organ. 15 (2000)
29. J.L. Sublett, J. Seltzer, R. Burkhead, P.B. Williams, H.J. Wedner, W. Phipatanakul, Air filters and air cleaners: rostrum by the American academy of allergy, asthma & immunology indoor allergen committee. J. Allergy Clin. Immunol. 125(1), 32–38 (2010). https://doi.org/10.1016/j.jaci.2009.08.036
30. J.R. Armstrong, H. Campbell, Indoor air pollution exposure and lower respiratory infections in young Gambian children. Int. J. Epidemiol. 20(2), 424–429 (1991)
31. W.J. Seow et al., Indoor concentrations of nitrogen dioxide and sulfur dioxide from burning solid fuels for cooking and heating in Yunnan Province, China. Indoor Air 26(5), 776–783 (2016). https://doi.org/10.1111/ina.12251
32. H.R. Anderson, Respiratory abnormalities and ventilatory capacity in a Papua New Guinea Island community. Am. Rev. Respir. Dis. 114(3), 537–548 (1976). https://doi.org/10.1164/arrd.1976.114.3.537
33. H.R. Anderson, Respiratory abnormalities, smoking habits and ventilatory capacity in a highland community in Papua New Guinea: prevalence and effect on mortality. Int. J. Epidemiol. 8(2), 127–136 (1979). https://doi.org/10.1093/ije/8.2.127
34. M.R. Pandey, Prevalence of chronic bronchitis in a rural community of the Hill Region of Nepal. Thorax 39(5), 331–336 (1984)
35. K.A. Qureshi, Domestic smoke pollution and prevalence of chronic bronchitis/asthma in a rural area of Kashmir. Indian J. Chest Dis. Allied Sci. 36(2), 61–72 (1994)
36. L.J. Rosen, V. Myers, J.P. Winickoff, J. Kott, Effectiveness of interventions to reduce tobacco smoke pollution in homes: a systematic review and meta-analysis. Int. J. Environ. Res. Publ. Health 12(12), 16043–16059 (2015). https://doi.org/10.3390/ijerph121215038
37. D.G. Fullerton, N. Bruce, S.B. Gordon, Indoor air pollution from biomass fuel smoke is a major health concern in the developing world. Trans. R. Soc. Trop. Med. Hyg. 102(9), 843–851 (2008). https://doi.org/10.1016/j.trstmh.2008.05.028
38. V. Mishra, Effect of indoor air pollution from biomass combustion on prevalence of asthma in the elderly. Environ. Health Perspect. 111(1), 71–78 (2003). https://doi.org/10.1289/ehp.5559
39. V.K. Mishra, R.D. Retherford, K.R. Smith, Biomass cooking fuels and prevalence of tuberculosis in India. Int. J. Infect. Dis. 3(3), 119–129 (1999)
40. M. Mohan et al., India-US case-control study of age-related cataracts. India-US case-control study group. Arch. Ophthalmol. 107(5), 670–676 (1989)
41. J.M. Tielsch et al., Exposure to indoor biomass fuel and tobacco smoke and risk of adverse reproductive outcomes, mortality, respiratory morbidity and growth among newborn infants in south India. Int. J. Epidemiol. 38(5), 1351–1363 (2009). https://doi.org/10.1093/ije/dyp286
42. D.V. Mavalankar, C.R. Trivedi, R.H. Gray, Levels and risk factors for perinatal mortality in Ahmedabad, India. Bull. World Health Organ. 69(4), 435–442 (1991)
43. O. US EPA, *Particulate Matter (PM) Basics*. US EPA, 19 Apr 2016. https://www.epa.gov/pm-pollution/particulate-matter-pm-basics. Accessed 25 Jan 2021
44. Y.-H. Cheng, Measuring indoor particulate matter concentrations and size distributions at different time periods to identify potential sources in an office building in Taipei City. Build. Environ. 123, 446–457 (2017). https://doi.org/10.1016/j.buildenv.2017.07.025
45. Z. Li, Q. Wen, R. Zhang, Sources, health effects and control strategies of indoor fine particulate matter (PM2.5): a review. Sci. Total Environ. 586, 610–622 (2017). https://doi.org/10.1016/j.scitotenv.2017.02.029
46. S. Baldacci et al., Allergy and asthma: effects of the exposure to particulate matter and biological allergens. Respir. Med. 109(9), 1089–1104 (2015). https://doi.org/10.1016/j.rmed.2015.05.017
47. R.D. Brook et al., Particulate matter air pollution and cardiovascular disease. Circulation 121(21), 2331–2378 (2010). https://doi.org/10.1161/CIR.0b013e3181dbece1
48. O. US EPA, *Carbon Monoxide's Impact on Indoor Air Quality*. US EPA, 31 July 2014. https://www.epa.gov/indoor-air-quality-iaq/carbon-monoxides-impact-indoor-air-quality. Accessed 25 Jan 2021

70. Q. Zhang, P.L. Jenkins, Evaluation of ozone emissions and exposures from consumer products and home appliances. Indoor Air **27**(2), 386–397 (2017). https://doi.org/10.1111/ina.12307

71. C. Guo, Z. Gao, J. Shen, Emission rates of indoor ozone emission devices: a literature review. Build. Environ. **158**, 302–318 (2019). https://doi.org/10.1016/j.buildenv.2019.05.024

72. World Health Organization. Regional Office for Europe, *Indoor Air Quality: Biological Contaminants: Report on a WHO Meeting, Rautavaara, 29 August–2 September 1988* (World Health Organization. Regional Office for Europe, 1990)

73. S. Capolongo, G. Settimo, Indoor air quality in healing environments: impacts of physical, chemical, and biological environmental factors on users, in *Indoor Air Quality in Healthcare Facilities*, ed. by S. Capolongo, G. Settimo, M. Gola (Springer International Publishing, Cham, 2017), pp. 1–11

74. R.C. Bruno, Sources of indoor radon in houses: a review. J. Air Pollut. Control Assoc. **33**(2), 105–109 (1983). https://doi.org/10.1080/00022470.1983.10465550

75. A. Ruano-Ravina, A. Fernández-Villar, J.M. Barros-Dios, Residential radon and risk of lung cancer in never-smokers. Arch. Bronconeumol. **53**(9), 475–476 (2017). https://doi.org/10.1016/j.arbres.2017.01.004

76. M. Al Jassim, R. Isaifan, A review on the sources and impacts of radon indoor air pollution. J. Environ. Toxicol. Stud. **2**(1) (2018). https://doi.org/10.16966/2576-6430.112

77. A. Kumar, R.P. Chauhan, M. Joshi, B.K. Sahoo, Modeling of indoor radon concentration from radon exhalation rates of building materials and validation through measurements. J. Environ. Radioact. **127**, 50–55 (2014). https://doi.org/10.1016/j.jenvrad.2013.10.004

78. I.M. Madany, M. Salim Akhter, O.A. Al Jowder, The correlations between heavy metals in residential indoor dust and outdoor street dust in Bahrain. Environ. Int. **20**(4), 483–492 (1994). https://doi.org/10.1016/0160-4120(94)90197-X

79. M.N. Rashed, Total and extractable heavy metals in indoor, outdoor and street dust from Aswan City, Egypt. Clean: Soil, Air, Water **36**(10–11), 850–857 (2008). https://doi.org/10.1002/clen.200800062

80. Y. Faiz, M. Tufail, M.T. Javed, M.M. Chaudhry, Naila-Siddique, Road dust pollution of Cd, Cu, Ni, Pb and Zn along Islamabad Expressway, Pakistan. Microchem. J. **92**(2), 186–192 (2009). https://doi.org/10.1016/j.microc.2009.03.009

81. T. Brown et al., Relationships between socioeconomic and lifestyle factors and indoor air quality in French dwellings. Environ. Res. **140**, 385–396 (2015). https://doi.org/10.1016/j.envres.2015.04.012

82. O.P. Kurmi, K.B.H. Lam, J.G. Ayres, Indoor air pollution and the lung in low- and medium-income countries. Eur. Respir. J. **40**(1), 239–254 (2012). https://doi.org/10.1183/09031936.00190211

83. S.J. Emmerich, K.Y. Teichman, A.K. Persily, Literature review on field study of ventilation and indoor air quality performance verification in high-performance commercial buildings in North America. Sci. Technol. Built Environ. **23**(7), 1159–1166 (2017). https://doi.org/10.1080/23744731.2016.1274627

84. I. Cretescu, D.N. Isopescu, D. Lutic, G. Soreanu, Indoor air pollutants and the future perspectives for living space design. Indoor Environ. Health (2019). https://doi.org/10.5772/intechopen.87309

85. L. Calderón-Garcidueñas et al., Long-term air pollution exposure is associated with neuroinflammation, an altered innate immune response, disruption of the blood-brain barrier, ultrafine particulate deposition, and accumulation of amyloid beta-42 and alpha-synuclein in children and young adults. Toxicol. Pathol. **36**(2), 289–310 (2008). https://doi.org/10.1177/0192623307313011

86. W.J. Fisk, A.H. Rosenfeld, Estimates of improved productivity and health from better indoor environments. Indoor Air **7**(3), 158–172 (1997). https://doi.org/10.1111/j.1600-0668.1997.t01-1-00002.x

87. X. Zhang, P. Wargocki, Z. Lian, C. Thyregod, Effects of exposure to carbon dioxide and bioeffluents on perceived air quality, self-assessed acute health symptoms, and cognitive performance. Indoor Air **27**(1), 47–64 (2017). https://doi.org/10.1111/ina.12284

88. S.B. Baker, W. Xiang, I. Atkinson, Internet of things for smart healthcare: technologies, challenges, and opportunities. IEEE Access **5**, 26521–26544 (2017). https://doi.org/10.1109/ACCESS.2017.2775180
89. Aditya, M. Sharma, S.C. Gupta, An internet of things based smart surveillance and monitoring system using Arduino, in *2018 International Conference on Advances in Computing and Communication Engineering (ICACCE)*, Jun 2018, pp. 428–433. https://doi.org/10.1109/ICACCE.2018.8441725
90. M. Fazio, A. Celesti, A. Puliafito, M. Villari, Big data storage in the cloud for smart environment monitoring. Proc. Comput. Sci. **52**, 500–506 (2015). https://doi.org/10.1016/j.procs.2015.05.023

Printed in the United States
by Baker & Taylor Publisher Services